Endoreversible Thermodynamics of Solar Energy Conversion

Endoreversible Thermodynamics of Solar Energy Conversion

Alexis De Vos

Universiteit Gent

Illustrations by

Jean Bekaert

OXFORD UNIVERSITY PRESS

1992

Oxford University Press, Walton Street, Oxford OX2 6DP
Oxford New York Toronto
Delhi Bombay Calcutta Madras Karachi
Petaling Jaya Singapore Hong Kong Tokyo
Nairobi Dar es Salaam Cape Town
Melbourne Auckland
and associated companies in
Berlin Ibadan

Oxford is a trade mark of Oxford University Press

Published in the United States
by Oxford University Press, New York

A catalogue record for this book is available from the British Library

Library of Congress Cataloging in Publication Data
Vos. Alexis de.
Endoreversible thermodynamics of solar energy conversion / Alexis de Vos; illustrations by Jean Bekaert.
Includes index.
1. Solar energy. 2. Thermodynamics. I. Title.
TJ810.V65 1992 621.47—dc20 92-13232
ISBN 0 19 851392 5

Text keyed by the author using LaTeX
Printed in Great Britain by
Bookcraft (Bath) Ltd
Midsomer Norton, Avon

C'est à la chaleur que doivent être attribués les grands mouvements qui frappent nos regards. C'est dans cet immense réservoir que nous pouvons puiser la force mouvante nécessaire à nos besoins.

Pour envisager dans toute sa généralité le principe de la production du mouvement par la chaleur, il faut le concevoir indépendamment d'aucun méchanisme, d'aucun agent particulier; il faut établir des raisonnements applicables à toute machine imaginable, quelle que soit la substance mise en œvre et quelle que soit la manière dont on agisse sur elle.

Sadi Carnot, 1824

Preface

This book attempts to describe various forms of solar energy conversion techniques in a unified way. The physical framework used to describe the various conversions is endoreversible thermodynamics, a recently developed subset of irreversible thermodynamics. It thus studies situations which are not in equilibrium and in which therefore entropy is continuously created. Nevertheless, the mathematics are simple because we consider only stationary situations; that is, we are not concerned with transient phenomena, oscillations, and so on.

Most undergraduate textbooks on thermodynamics emphasize equilibrium thermodynamics and reversible processes. No entropy is created and conversion efficiencies are maximal: equal to the Carnot efficiency. For irreversible conversion processes, the reader learns only that entropy production is positive and that conversion efficiency is lower than the Carnot efficiency. But how great the entropy creation is, and how low the efficiency, is usually not expressed. Endoreversible thermodynamics give us the opportunity to calculate explicit values for a broad class of processes. It is demonstrated in the text that solar energy conversion is a process particularly well suited to being described in this way.

The book is not a textbook, nor a reference manual. It does not aim at completeness: it neither describes all conversion processes nor gives a complete state-of-the-art description of the technology. On the contrary, it illustrates general principles using idealized models. Technological examples are presented only to compare reality with theory.

Solar energy specialists will recognize their everyday technology, placed in a general framework. Non-specialists will discover a wide overview of solar energy conversion.

The language of the book is fairly mathematical. However, the mathematical knowledge of anyone who has followed at least two years of a course on mathematics, physics, or chemistry at university level should amply suffice in order to follow the calculations: the book does not contain a single differential equation! Also, the physics background needed is of undergraduate level. I therefore hope the book will appeal to a broad public. However, the reader should be familiar with SI units, which are used throughout the book (except for the electronvolt and some rare appearances of the degree centigrade).

Sometimes the physics (or sometimes the mathematics) background is more deeply explored. Such sections can be skipped by readers who are interested only in pure solar energy results. *Gourmet* readers will enjoy excursions in the sections labeled with the symbol ♡.

I wish to thank especially the two people whose collaboration has been fruitful for many years and without whom the present book would never have been written:

- Prof. H. Pauwels of the Universiteit Gent and
- Prof. P. T. Landsberg of the University of Southampton.

I also wish to thank the many people whose comments and suggestions have been a continuing encouragement:

- Prof. M. Vanwormhoudt and Prof. C. Grosjean of the Universiteit Gent,
- Prof. P. Baruch of the Université Paris 7,
- Prof. J. Parrott of the University of Wales Institute of Science and Technology, and
- Prof. W. Ruppel and Dr P. Würfel of the Universität Karlsruhe.

The following institutions gave aid and comfort:

- the *v.z.w.* Interuniversitair Micro Elektronika Centrum,
- the Commission of European Communities through the Joule grant JOUR-0066, and
- the `CAGe` group of the Universiteit Gent for help with the computer algebra systems `Reduce` and `Maple`.

The following people are thanked for their direct help with the completion of the text:

- J. Bekaert for drawing all figures, i.e. the scientific and technical diagrams as well as the artist's impressions,
- students J. Landries, P. van der Wel, and D. Vyncke, for performing part of the numerical computations,
- Prof. E. Selleslagh (Universiteit Gent), Prof. G. De Mey (Universiteit Gent), Dr J. Gretz (Centro Commune di Ricercha, Ispra), Dr B. Aubert (Dynatech Automation, Saint-Quentin en Yvelines), Prof. W. Bloss (Universität Stuttgart), and Prof. J. Loferski (Brown University, Providence) for providing valuable information for the examples,

- M. Deschrijver for providing the computer aids, especially the LaTeX facility,
- E. De Ley for carefully screening the whole text, and
- The staff of Oxford University Press for their help in completion of the book.

Gent
March 1992

A. De V.

Contents

1
Radiation

1.1 Introduction

For the study of solar energy conversion, one needs to know how the photon flux of radiation is distributed over the different photon energies E. The calculation of the spectral distribution $n(E)$ is supported by two distinct pillars:

- $h(E)dE$, i.e. the number of photon modes with energy in the range $(E, E + dE)$, and
- $f(E)$, i.e. the occupation probability of a mode with energy E.

Although we could postulate the functions $h(E)$ and $f(E)$, we will derive them from first principles. This is covered in Sections 1.2 and 1.3. Then, in Section 1.4, we will use these results to derive the radiation spectrum $n(E)$. Finally, in Section 1.5, we will see how $n(E)$ yields the total photon flux

$$N = \int_0^\infty n(E)\, dE \ ,$$

as well as the associated energy flux

$$Q = \int_0^\infty n(E) E\, dE \ .$$

Any reader who whishes to skip the calculations of $h(E)$ and $f(E)$ can proceed immediately to Section 1.4.

1.2 Photon modes ♡

We consider three-dimensional position space (x_1, x_2, x_3) and the corresponding three-dimensional momentum space (p_1, p_2, p_3). Together they form the six-dimensional hyperspace $(x_1, x_2, x_3, p_1, p_2, p_3)$, called the phase

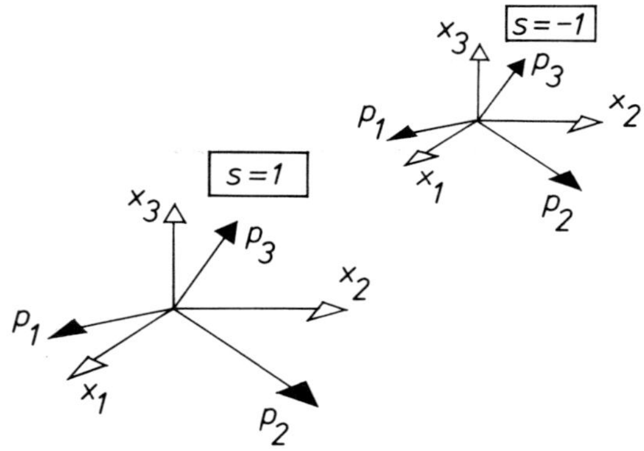

Fig. 1.1. Photon phase space.

space. The three position coordinates x_i of a photon can have any value, ranging from $-\infty$ to $+\infty$, as can the three momentum coordinates p_i.

Because a photon is a spin 1 particle, it has an additional coordinate, called the spin s. In contrast with the other six coordinates, the spin coordinate can have only two discrete values: -1 and $+1$. Therefore the complete phase space of the photon consists of two six-dimensional hyperspaces, one with $s = -1$ (spin down) and one with $s = 1$ (spin up). See Fig. 1.1.

Let dg be the number of modes with first position coordinate in the range $(x_1, x_1 + dx_1)$, with second position coordinate in the range $(x_2, x_2 + dx_2)$, ..., and third momentum coordinate in the range $(p_3, p_3 + dp_3)$. These modes occupy one of the two elemental hypervolumes

$$dx_1.dx_2.dx_3.dp_1.dp_2.dp_3 \ . \tag{1.1}$$

After the Heisenberg principle of quantum mechanics one particle mode occupies a volume h^3. Here h is a fundamental constant of nature, i.e. Planck's constant 6.63×10^{-34} Js. Therefore dg equals the number of times h^3 fits into the two volumes (1.1):

$$dg = \frac{2\ dx_1.dx_2.dx_3.dp_1.dp_2.dp_3}{h^3} \ . \tag{1.2}$$

Let g be the number of modes with momentum

$$\sqrt{p_1^2 + p_2^2 + p_3^2}$$

smaller than or equal to p. By simply integrating eqn (1.2) with respect to the three space coordinates over the volume V and with respect to the three momentum coordinates over a sphere with radius p, we get

$$g = \frac{2V\frac{4}{3}\pi p^3}{h^3} \ . \tag{1.3}$$

Here π is the mathematical constant, simply called pi, with the well-known numerical value 3.14159.

Because photons are particles with zero rest mass, they always travel at the same speed c, called the speed of light, i.e. 300 Mm/s, a second fundamental constant of nature. As a consequence, the energy E and the momentum p of the photon are related by the simple relativistic equation

$$E = cp \ .$$

The number of photons with energy equal to or smaller than E is thus

$$g = \frac{\frac{8}{3}\pi V E^3}{c^3 h^3} \ . \tag{1.4}$$

By differentiation with respect to E, we find the number of modes per unit of energy:

$$\frac{dg}{dE} = \frac{8\pi V E^2}{c^3 h^3} \ . \tag{1.5}$$

Having obtained the number of modes in the volume V, we now need the number of modes leaving this volume through a surface area S. See Fig. 1.2. Since radiation in a cavity is isotropic, i.e. uniformly distributed over all directions, a fraction $d\omega/4\pi$ has a propagation direction in the elemental solid angle $d\omega$. The photons with such a direction, hitting the surface area S in a time interval dt, occupy a cylindrical volume $V = Sc\cos\vartheta\, dt$, where ϑ is the angle between the pencil $d\omega$ and the normal on S. Thus eqn (1.5) becomes

$$\begin{aligned} \frac{dg}{dE} &= \frac{8\pi(Sc\cos\vartheta\, dt)(d\omega/4\pi)E^2}{c^3 h^3} \\ &= \frac{2E^2}{c^2 h^3}\, Sdt\cos\vartheta\, d\omega \ . \end{aligned} \tag{1.6}$$

Photons hit S from a hemisphere, i.e. from a solid angle 2π. Integration of eqn (1.6) over this hemisphere yields

$$\frac{dg}{dE} = \frac{2\pi E^2}{c^2 h^3}\, Sdt \ ,$$

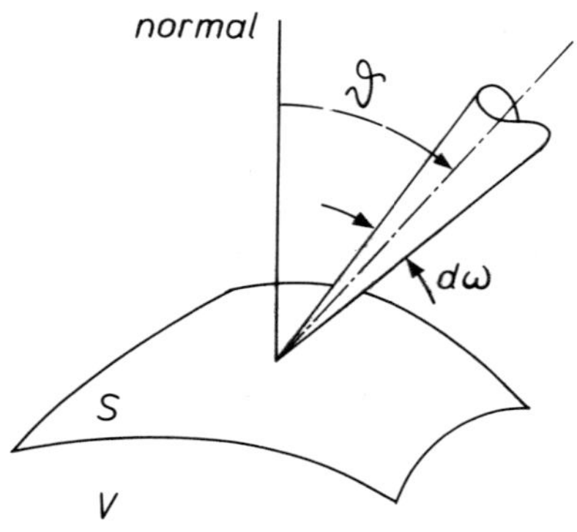

Fig. 1.2. Photon flux through a surface area S.

because of Lambert's geometrical law:

$$\iint_{2\pi} \cos\vartheta \, d\omega = \pi \; .$$

If we call $h(E)$ the number of photons hitting the boundary surface of the volume V per unit energy interval, per unit time and per unit surface area, we eventually have

$$h(E) = \frac{2\pi}{c^2 h^3} \, E^2 \; . \tag{1.7}$$

The number of photons leaving the volume V along the other side of the surface is then given by

$$\epsilon(E) h(E) = \epsilon(E) \, \frac{2\pi}{c^2 h^3} \, E^2 \; ,$$

where ϵ is called the emissivity of the surface. The fraction $[1 - \epsilon(E)] \, h(E)$ of the incident flux $h(E)$ is reflected by the surface S and thus sent back to the inside of the cavity volume V.

1.3 Photon statistics ♡

As photons do not interact with each other, their statistical properties are entirely governed by their interaction with matter. Therefore we necessarily need to know something more about the statistics of matter. Because we

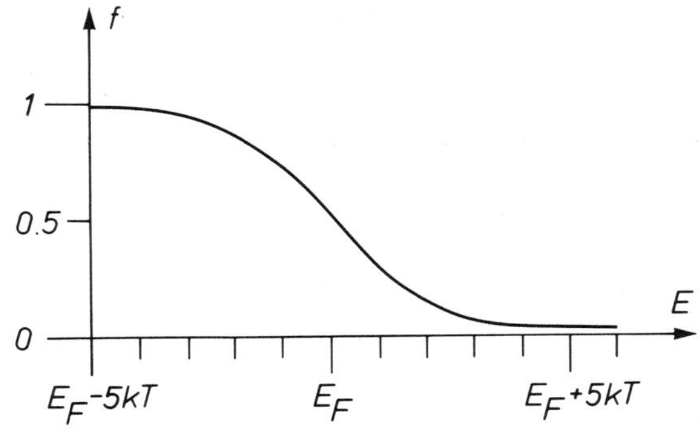

Fig. 1.3. The Fermi function.

cannot reinvent all laws of physics from scratch, we will postulate some fundamental statistical laws of the solid state, without any proof.

The probability of a particular energy level with energy E to be occupied by an electron is

$$f(E) = \frac{1}{\exp(\frac{E-E_F}{kT}) + 1} , \tag{1.8}$$

where E_F is the Fermi energy and T the temperature associated with the matter reservoir, whereas k is a fundamental constant, i.e. the Boltzmann constant or 1.38×10^{-23} J/K. Up to now, we have introduced three fundamental constants of nature: h, c, and k. Figure 1.3 shows the function $f(E)$. We see that $f \approx 1$ for $E \ll E_F$, that $f = 1/2$ for $E = E_F$, and that $f \approx 0$ for $E \gg E_F$.

Let us consider two states of matter, labelled 1 and 2, respectively. Let state 2 be the one with the higher energy:

$$E_2 > E_1 .$$

We consider photons with energy E equal to the energy difference

$$E_2 - E_1 .$$

See Fig. 1.4. According to the Einstein model, these photons can interact with the matter in three different ways:

Stimulated absorption. An existing photon is absorbed by an electron at level 1, causing the electron to jump to level 2. The probability of the event is proportional to:

- f, i.e. the probability that there is a photon,
- f_1, i.e. the probability that level 1 is occupied, and
- $(1 - f_2)$, i.e. the probability that level 2 is not occupied.

Spontaneous emission. An electron falls spontaneously from level 2 to level 1 by emitting a photon. The probability of the event is proportional to:

- f_2, i.e. the probability that level 2 is occupied, and
- $(1 - f_1)$, i.e. the probability that level 1 is not occupied.

Stimulated emission. Under influence of an already existing photon, an electron falls from level 2 to level 1 with emission of a second photon. The probability of the event is proportional to:

- f, i.e. the probability that there already is a photon,
- f_2, i.e. the probability that level 2 is occupied, and
- $(1 - f_1)$, i.e. the probability that level 1 is not occupied.

This interaction model leads to the following kinetic equation:

$$\frac{df}{dt} = a[\ -ff_1(1-f_2) + f_2(1-f_1) + ff_2(1-f_1)\ ]\ .$$

In steady-state conditions $df/dt = 0$, so that

$$-ff_1(1-f_2) + f_2(1-f_1) + ff_2(1-f_1) = 0$$

or

$$f = \frac{1}{\frac{f_1(1-f_2)}{(1-f_1)f_2} - 1}\ . \tag{1.9}$$

After eqn (1.8), we assume that

$$f_1 = \frac{1}{\exp(\frac{E_1 - E_{F1}}{kT}) + 1} \tag{1.10}$$

$$f_2 = \frac{1}{\exp(\frac{E_2 - E_{F2}}{kT}) + 1}\ , \tag{1.11}$$

where we deliberately assumed that the two energy levels are at the same temperature T, but do not necessarily have the same Fermi energy E_F. By substitution of eqns (1.10) and (1.11) into eqn (1.9), we immediately find

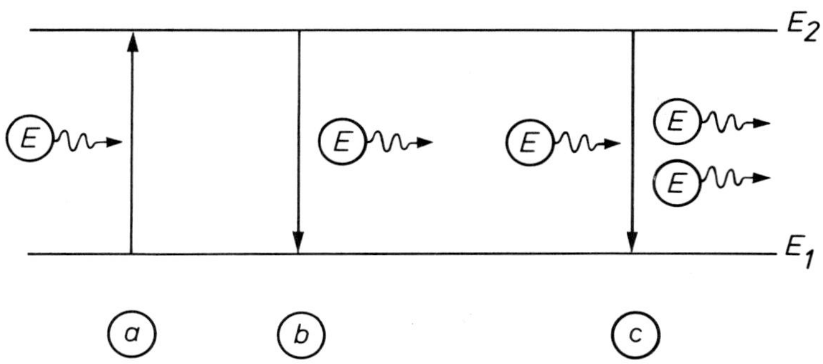

Fig. 1.4. Interaction of photons with two energy states of matter: (a) stimulated absorption, (b) spontaneous emission, (c) stimulated emission.

$$f = \frac{1}{\exp(\frac{E_2 - E_1 - E_{F2} + E_{F1}}{kT}) - 1}$$

or

$$f(E) = \frac{1}{\exp(\frac{E-\mu}{kT}) - 1} , \tag{1.12}$$

where

$$\mu = E_{F2} - E_{F1}$$

will be called the chemical potential of the photons. Equation (1.12) is called the Bose statistics of photons. Figure 1.5 shows the Bose function $f(E)$. We see that $f \to +\infty$ for $E \to \mu$ and $f \to 0$ for $E \to +\infty$. For $E < \mu$, eqn (1.12) gives negative values, but makes no physical sense: for $E < \mu$, we actually have $f = 0$.

1.4 Planck's law

The number of photons n in an energy range $(E, E + dE)$ is given by the product of the number of photon modes in the range and the occupation probability of these modes:

$$n(E) = \epsilon(E) h(E) f(E) .$$

With the expression for $h(E)$ found in Section 1.2 and the expression for $f(E)$ found in Section 1.3, we get

$$n(E) = \epsilon(E) \, \frac{2\pi}{c^2 h^3} \, \frac{E^2}{\exp(\frac{E-\mu}{kT}) - 1} .$$

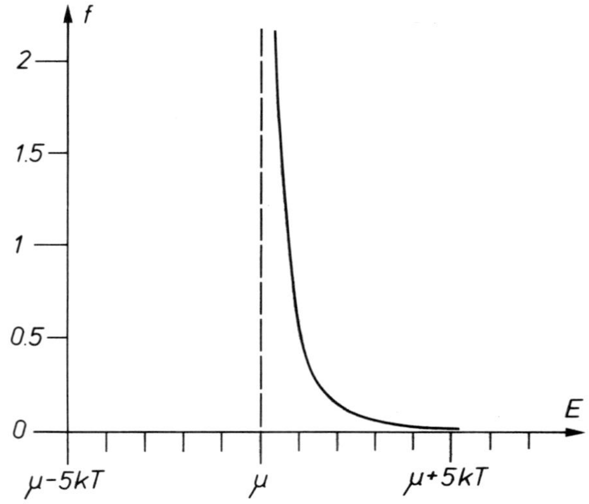

Fig. 1.5. The Bose function.

This important law is called Planck's law. Because a distribution over energies is often called a spectrum, we will call it Planck's spectrum.

The most well-known case is the radiation spectrum of a reservoir which has an emissivity ϵ independent of energy and which has a zero chemical potential μ:

$$\begin{aligned}\epsilon(E) &= \epsilon \\ \mu &= 0\,.\end{aligned}$$

It is called the grey-body spectrum:

$$n(E) = \epsilon\,\frac{2\pi}{c^2h^3}\,\frac{E^2}{\exp(\frac{E}{kT})-1}\,. \qquad (1.13)$$

The special case where ϵ equals unity is called the black-body spectrum. Figure 1.6 shows the black-body function $n(E)$. For small E, i.e. for $E \ll kT$, we have $\exp(E/kT) \approx 1 + (E/kT)$, so that $n(E)$ is proportional to E. For large E, i.e. for $E \gg kT$, $n(E)$ goes to zero like $E^2\exp(-E/kT)$. For some intermediate value, called Wien's energy, the curve displays a maximum. Calculation of Wien's energy involves the numerical solution of a transcendental equation, yielding $E = 1.59362\ kT$. Thus the abscissa value of the maximum changes proportionally with temperature. This fact

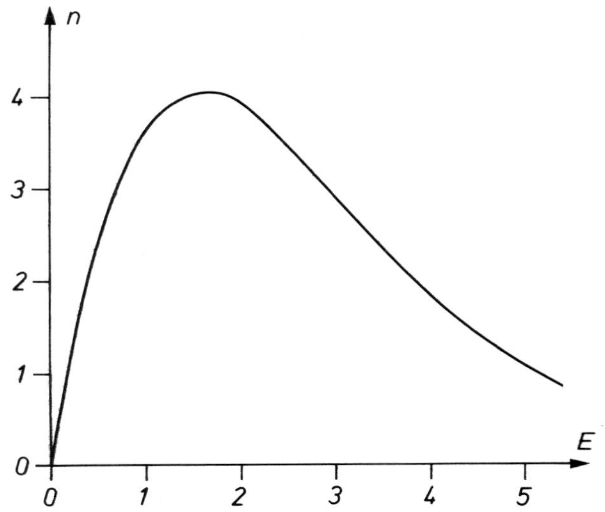

Fig. 1.6. The Planck spectrum with linear scales: E in units kT; n in units k^2T^2/c^2h^3.

is called Wien's displacement law. The ordinate value of the maximum changes even more dramatically with temperature T. As a consequence we have to use logarithmic scales, if we want to plot different spectra for different temperatures on a single figure. An example is shown in Fig. 1.7. The three values chosen for T, i.e. 2.7, 288, and 5762 K, seem somewhat arbitrary. The reason for this choice will, however, become apparent in Chapter 2.

As we can see from Fig. 1.7, all three temperature examples give rise to spectra concentrated around energies, which are very small when expressed in the SI unit of energy, i.e. the joule. It is therefore advantageous to introduce a technical energy unit, i.e. the electronvolt. One electronvolt is defined as the energy of the elementary charge q at a voltage of 1 volt. We have

$$1 \text{ eV} = 1.60 \times 10^{-19} \text{ J},$$

as q equals 1.60×10^{-19} coulomb. Note that q is the fourth (and last) fundamental constant of nature that we introduce in the present book. The three other constants were: h, c, and k. See also Appendix A.

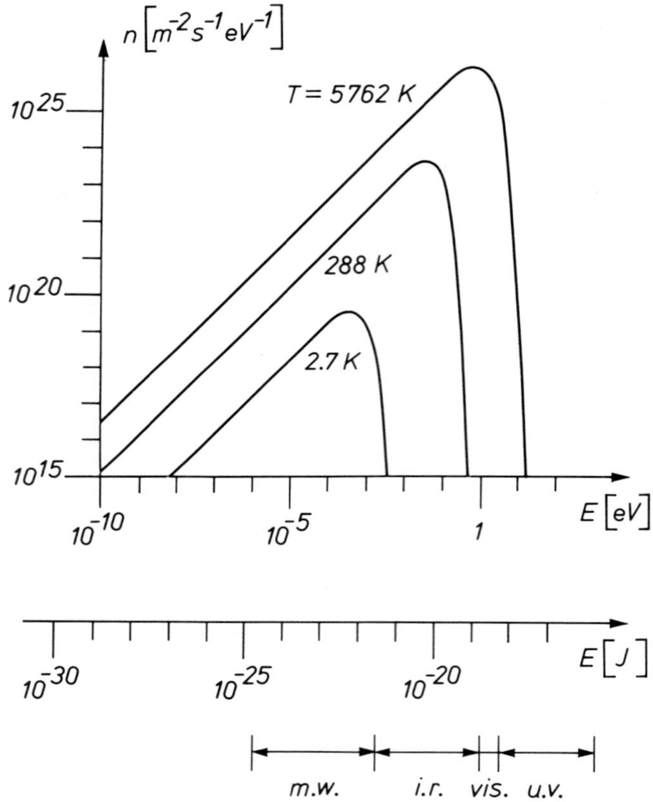

Fig. 1.7. The Planck spectrum with logarithmic scales (m.w., micro waves; i.r., infrared radiation; vis., visible light; u.v., ultraviolet radiation).

1.5 The Stefan–Boltzmann law

By integrating the grey-body spectrum over all energies E, we obtain the total flux of photons (per unit time and per unit surface area):

$$\begin{aligned} N &= \int_0^\infty n(E)\,dE \\ &= \epsilon\,\frac{2\pi}{c^2h^3}\,(kT)^3 \int_0^\infty \frac{x^2\,dx}{\exp(x)-1}\,. \end{aligned} \tag{1.14}$$

The definite integral at the end of this expression is written as $2\,\zeta(3)$ by mathematicians. Indeed, they define the Riemann Zeta function $\zeta(n)$ by

$$(n-1)!\;\zeta(n) = \int_0^{+\infty} \frac{x^{n-1}\,dx}{\exp(x)-1}\,.$$

The value of $\zeta(n)$ is tabulated and in the table we can find that $\zeta(3)$ equals about 1.20206. See Appendix A. Thus we finally find the number of photons emitted by a grey body:

$$N = 1.20206\;\epsilon\;\frac{4\pi k^3}{c^2h^3}\;T^3.$$

We can also calculate the energy emitted by the body. For that purpose we have to calculate the integral

$$\begin{aligned} Q &= \int_0^\infty E\,n(E)\,dE \\ &= \epsilon\,\frac{2\pi}{c^2h^3}\,(kT)^4 \int_0^\infty \frac{x^3\,dx}{\exp(x)-1} \\ &= \epsilon\,\frac{2\pi}{c^2h^3}\,(kT)^4\;6\;\zeta(4)\,. \end{aligned} \tag{1.15}$$

In contrast to $\zeta(3)$, the value of $\zeta(4)$ is not only to be found in the tables, it can also be reduced to a 'simpler' mathematical number, i.e. π:

$$\zeta(4) = \frac{\pi^4}{90} = 1.08232\,.$$

Therefore

$$Q = \epsilon\;\frac{2\pi^5k^4}{15c^2h^3}\;T^4\,.$$

This expression is so important, that it has a name of its own: the Stefan–Boltzmann equation. The coefficient of T^4 has also a name: the Stefan–Boltzmann constant, with notation σ. The latter is not a fundamental constant of nature, as it can be expressed in terms of the three fundamental numbers h, c, and k:

$$\sigma = \frac{2\pi^5 k^4}{15c^2h^3} \ .$$

Its value can thus be calculated and amounts to 5.67×10^{-8} $\mathrm{Wm^{-2}K^{-4}}$.

Thus we can write

$$Q = \epsilon\,\sigma T^4 \ .$$

Analogously, we can write

$$N = \epsilon\,\sigma' T^3 \ ,$$

where σ' can also be expressed in terms of h, c, and k:

$$\sigma' = \frac{4\pi\zeta(3)k^3}{c^2h^3}$$

and has a value of 1.51×10^{15} $\mathrm{s^{-1}m^{-2}K^{-3}}$. Oddly enough, σ' did not receive any name.

We end this section by noting that the average photon energy of grey (and of black) body radiation equals

$$\frac{Q}{N} = \frac{\sigma}{\sigma'}\,T = \frac{\pi^4}{30\zeta(3)}\,kT = 2.70117\ kT \ .$$

This number is somewhat larger than Wien's energy, $1.59362\ kT$. In the language of statistics we would say that the Planck distribution has a mean value larger than the mode.

1.6 Kirchhoff's law

Until now, we have been concerned with the light emitted by a body. We also need to know something about the light absorbed by it. A black body absorbs all light incident on it. A non-black body absorbs a fraction $\alpha(E)$ of the light incident on it, where α is called the absorptivity of the body. We now give, without proof, an important law of radiation physics, Kirchhoff's law, which states that the absorptivity exactly equals the emissivity:

$$\alpha(E) = \epsilon(E) \ .$$

In particular, a grey body with emissivity ϵ will absorb exactly the fraction ϵ of all light incident on it.

1.7 Why T^4 ? ♡

As a result of the Stefan–Boltzmann law and the Kirchhoff law, two black surfaces facing each other exchange a net heat flux:

$$Q = \sigma S(T_1^4 - T_2^4) \ ,$$

where S is the surface area (assumed equal for the two surfaces) and where T_1 and T_2 are the temperatures of the bodies.

Why is radiation exchange governed by a fourth-power law? After all, heat conduction is satisfying a simpler equation:

$$Q = \frac{\lambda S}{d}(T_1 - T_2) \ ,$$

where λ is the heat conductivity of the medium between the two reservoirs and where d is the distance between the two surfaces. See Fig. 1.8. This law, sometimes called Fourier's law and sometimes called Newton's law, is linear like many so-called phenomenological or constitutive laws in physics. Ohm's law of electricity analogously states that the electrical current is proportional to the voltage difference $V_1 - V_2$. Fick's law of particle diffusion analogously states that the particle current is proportional to the concentration difference $n_1 - n_2$. Poisseuille's law says that the volume rate of a laminar fluid flow is proportional to the pressure difference $p_1 - p_2$. Etc.

So, why is the transport law of Stefan–Boltzmann not of this general type? Let us retrace the calculations of Sections 1.2 - 1.4 to find the answer. The T^4 proportionality appears in eqn (1.15), where it is introduced by replacing $E^3 dE$ by $(kT)^4 x^3 dx$. Thus the T^4 is a consequence of the factor E^3 under the integral. This E^3 factor comes from $E\, n(E)$. Thus the T^4 is a consequence of $n(E)$ containing a factor E^2 in eqn (1.13). This fact, in turn, is a consequence of $h(E)$ being proportional to E^2 in eqn (1.7) and thus of g being proportional to E^3 in eqn (1.4) or to p^3 in eqn (1.3). So, we eventually retrieved the origin of the T^4 behaviour of the Stefan–Boltzmann law: the volume of a sphere with radius p equals $\frac{4}{3}\pi p^3$. The volume of a sphere being proportional to the third power of the radius is a consequence of space being three-dimensional. Indeed, the volume of an n-dimensional hypersphere of radius p is proportional to p^n ; it equals

$$V_n \, p^n \ ,$$

where V_n denotes a constant, independent of p. In fact V_n is the volume of the n-dimensional hypersphere with radius 1. Its value is dependent only

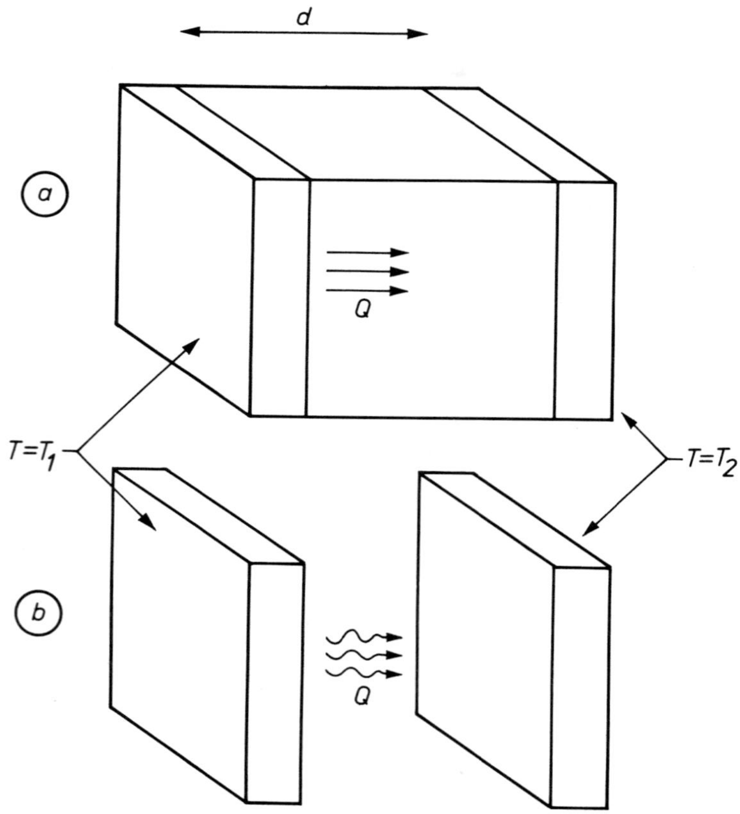

Fig. 1.8. Heat exchange by two surfaces: (a) by conduction, (b) by radiation.

on n. For $n = 3$, we have $V_3 = \frac{4}{3}\pi$. For other n, the value of V_n is tabulated in Appendix B.

If the universe were of n, rather than three dimensions, g would be proportional to E^n and $h(E)$ to E^{n-1}, and so $n(E)$ would contain a factor E^{n-1}. This would finally lead to a photon flux N proportional to T^n and a heat flux Q proportional to T^{n+1}. In other words, the power of T in the Stefan–Boltzmann law equals the number of space dimensions plus one. Explicit calculations [3] lead to the n-dimensional Stefan–Boltzmann law:

$$Q = \sigma_n T^{n+1} ,$$

where σ_n, i.e. the n-dimensional Stefan–Boltzmann constant, is given by

$$\sigma_n = 2\ \pi^{(n-1)/2}\ \frac{\Gamma(n+1)\zeta(n+1)}{\Gamma(\frac{n+1}{2})}\ \frac{k^{n+1}}{h^n c^{n-1}} ,$$

where $\Gamma(x)$ denotes the mathematical function called the gamma function. See Appendix B.

One can go one step further: why has our universe three spatial dimensions? The fact is: nobody in the world really knows. It is (still ...) one of the great mysteries of nature. Did space always have three dimensions? Superstring theories suggest it did not. Shortly after its Big Bang birth, the universe could have had nine or even 25 spatial dimensions. In those early times, radiation was thus proportional to T^{10} or even T^{26}...

Analogous results can be formulated for the particle flux N. We have in n dimensions

$$N = \sigma'_n T^n ,$$

with

$$\sigma'_n = \frac{\zeta(n)}{n\zeta(n+1)}\ \frac{\sigma_n}{k} .$$

We can only guess what the consequences of these facts would have been. Wisely, the remainder of this book will consider mostly three-dimensional space. We will make only about three excursions into hyperspace, just for fun.

Bibliography

1. A. De Vos (1982). Fundamental diode-noise. *Physica*, volume 113 A, pp. 235 - 252.

2. P. Würfel (1982). The chemical potential of radiation. *Journal of Physics C: Solid State Physics*, volume 15, pp. 3967 - 3985.

3. P. T. Landsberg and A. De Vos (1989). The Stefan–Boltzmann constant in n - dimensional space. *Journal of Physics A: Mathematical and General*, volume 22, pp. 1073 - 1084.

2
The solar system

2.1 The Sun

In order to determine the properties of solar radiation, we need to make measurements from an aircraft or a satellite, so that all influences of the Earth's atmosphere are eliminated. Figure 2.1 shows the energy's spectral distribution $En(E)$, obtained from photospectrometers in an aircraft 12 km above the Earth [1]. It closely resembles a Planck spectrum with zero chemical potential. If we search for the parameter value T, such that the spectrum

$$A \frac{E^3}{\exp(\frac{E}{kT_s}) - 1} \tag{2.1}$$

approximates to the experimental spectrum $En(E)$ as closely as possible, we find

$$T_s = 5762 \text{ K} .$$

Figure 2.1 demonstrates that the curve fitting is remarkably good. Thus we are allowed to say that the Sun is a body at 5762 K.

However, the value of A in eqn (2.1), used in the above curve fitting, is much smaller than the theoretical factor

$$\frac{2\pi}{c^2 h^3} ,$$

which we expect from eqn (1.13). Does this mean the Sun is a grey body with very small emissivity ϵ? No: the solar radiation reaching the Earth is much weaker than the solar radiation leaving the Sun. If 1 m^2 of the Sun's surface emits a flux Q_s, then the total flux emitted by the Sun is $4\pi R_s^2 Q_s$, with R_s the radius of the Sun. When the Sun's rays reach the Earth, this energy is no longer distributed over a surface area $4\pi R_s^2$, but over the much larger surface area $4\pi r^2$, with r the radius of the Earth's orbit around the

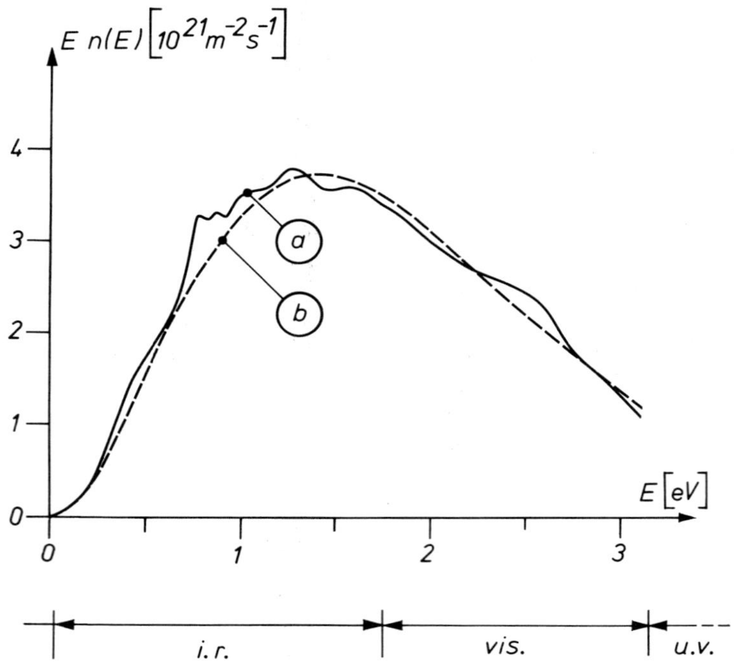

Fig. 2.1. The solar spectrum with linear scales: (a) experimental spectrum, (b) Planck spectrum (i.r., infrared radiation; vis., visible light; u.v., ultraviolet radiation).

Sun. Thus, because of the radial expansion of the Sun's rays, the energy density is now

$$\frac{4\pi R_s^2 Q_s}{4\pi r^2} = \frac{R_s^2}{r^2} Q_s \ .$$

The flux density is therefore diluted by a factor

$$f = \frac{R_s^2}{r^2} \ . \tag{2.2}$$

We have

$$\begin{aligned} R_s &= 696 \text{ Mm} \ , \\ r &= 150 \text{ Gm} \\ \text{and thus} \quad f &= 2.16 \times 10^{-5} \ . \end{aligned}$$

It turns out that A exactly satisfies

$$A = f\,\frac{2\pi}{c^2 h^3}\ ,$$

so that we can conclude that the Sun is in fact a black body, i.e. a body with $\epsilon = 1$.

The total irradiation of the upper atmosphere of the Earth by the Sun is called the solar constant S. We have

$$S = f\sigma T_s^4\ ,$$

which amounts to 1353 W/m^2.

The corresponding particle flux, i.e. photon flux, is given by

$$S' = f\sigma' T_s^3$$

and amounts to 6.24×10^{21} s^{-1}m^{-2}. The average energy of one photon is given by

$$\frac{S}{S'} = \frac{\sigma}{\sigma'}\,T_s = 2.70117\ kT_s = 1.34\ \text{eV}\ .$$

In fact, the term solar 'constant' for S is somewhat misleading. Indeed, the solar flux incident on the Earth varies as a function of time, because of two phenomena:

- The orbit of the Earth around the Sun is no circle with radius r. According to Kepler's first law, it is an ellipse with half major axis a and half minor axis b. Therefore the distance Sun–Earth varies periodically, with a period τ, given by

  $$\tau = 31.6\ \text{Ms}\ ,$$

  a time lapse we commonly call 1 year. When the Earth is at its perihelion,

  $$r = r_{min} = 147\ \text{Gm}$$

  and when the Earth is at its aphelion,

  $$r = r_{max} = 152\ \text{Gm}\ .$$

  Because solar flux is proportional to r^{-2}, this causes a 7 per cent difference between these two days. In order to avoid such complication, the solar constant is defined for the time average value of r^{-2}. Because of Kepler's second law, we can easily prove that the latter equals $1/ab$. See Appendix C. In other words: for irradiance, we may regard the Earth as a planet in a circular orbit around the Sun, with radius equal to $\sqrt{ab}$.

- The temperature T_s of the Sun is not a rigorous constant. It varies periodically, with a period τ which is not exactly known, but amounts to about 11 years, called the sunspot cycle. The amplitude of the oscillation is about 0.1 K, i.e. about 0.02 per cent. The corresponding relative variation of the solar constant S is four times larger, because of the T_s^4 dependence and thus about 0.1 per cent, i.e. about 1 W/m^2.

In the remainder of the book, we will consider T_s as a constant (5762 K). We will also consider the distance r as a constant, equal to $\sqrt{ab}$. We will apply such circular approximation to the orbits of all nine planets of the solar system.

2.2 The planets

Table 2.1 gives the value of $r = \sqrt{ab}$ for the nine planets of our solar system, as well as the corresponding values of the dilution factor $f = R_s^2/r^2$ and the solar constant $S = f\sigma T_s^4$.

We will now evaluate the thermal balance of the planets. In this chapter, the planet's surface will be assumed to be at a uniform temperature T_p.

We denote by R the radius of the planet. In this section, we will assume that the planet temperature T_p arises from the thermal balance between two radiations:

- the radiation $4\pi R^2 \sigma T_p^4$ arising from the planet's own radiation density σT_p^4, multiplied by the sphere surface area $4\pi R^2$, and
- the irradiation $\pi R^2 f \sigma T_s^4$ resulting from the solar radiation density $f\sigma T_s^4$, multiplied by the circle surface area πR^2, i.e. the surface area of the projection of the planet on a plane, perpendicular to the Sun's rays.

One can call $4\pi R^2$ the radiation cross-section of the planet, functioning as a transmitting antenna, whereas πR^2 is the capture cross-section of the planet, acting as a receiving antenna.

Thus the total incoming flux is

$$Q = 4\pi R^2 \sigma \left(\frac{f}{4} T_s^4 - T_p^4 \right) . \tag{2.3}$$

Now, thermal balance requires $Q = 0$ and thus eqn (2.3) becomes

$$\frac{f}{4} T_s^4 - T_p^4 = 0 , \tag{2.4}$$

yielding immediately

Table 2.1. Time-averaged distance of the various planets from the Sun and related quantities.

Planet	$\sqrt{ab}$ (Gm)	f	S (W/m^2)	T_p (K)	T_{exp} (K)
Mercury	57	1.48×10^{-4}	9228	449	440
Venus	108	4.14×10^{-5}	2586	327	733
Earth	150	2.16×10^{-5}	1353	278	288
Mars	227	9.37×10^{-6}	586	225	220
Jupiter	778	8.01×10^{-7}	50	122	124 (*)
Saturn	1426	2.38×10^{-7}	15	90	95 (*)
Uranus	2868	5.89×10^{-8}	4	64	59 (*)
Neptune	4497	2.40×10^{-8}	2	51	59 (*)
Pluto	5806	1.44×10^{-8}	1	45	≈ 50

(*) The H_2 atmospheres of the planets Jupiter, Saturn, Uranus, and Neptune are too dense to allow measurements of their surface; the indicated temperature T_{exp} is therefore the temperature of the upper cloud layers.

$$T_p = \frac{f^{1/4}}{\sqrt{2}} \, T_s \, . \tag{2.5}$$

Calculation of this expression yields the numbers T_p given in Table 2.1. We can compare these theoretical temperatures with the experimental surface temperatures T_{exp} actually measured.

In spite of the simple model, we see a good agreement between theoretical and experimental values. Only the high surface temperature of Venus cannot be explained by the simple theory. The high temperature of the surface of Venus is caused by the pronounced greenhouse effect of its dense CO_2 atmosphere.

Secondary effects can be taken into account by introducing an emissivity $\epsilon(E)$ of the planet. We assume that $\epsilon(E)$ has a constant value $1-\gamma$ in the far-infrared spectrum (small E) and another, but also constant, value $1-\rho$ in the near-infrared and visible part of the electromagnetic spectrum (large E). The emitted flux density σT_p^4 is thus attenuated by a factor $1-\gamma$ and only a fraction $1-\rho$ of the incident flux density $f\sigma T_s^4$ is absorbed. Therefore the net incoming flux, i.e. absorption minus emission, is given by

$$Q = 4\pi R^2 \sigma \, [\, (1-\rho)\frac{f}{4} T_s^4 - (1-\gamma) T_p^4 \,] \, .$$

Therefore the radiation balance of eqn (2.4) now becomes

Table 2.2. Albedo and greenhouse factor of the various planets and related quantities.

Planet	ρ	γ	Z (W/m^2)	T_p (K)	T_{exp} (K)
Mercury	0.1	0.0	8306	439	440
Venus	0.7	0.99	776	765	733
Earth	0.3	0.4	947	289	288
Mars	0.2	0.2	468	225	220
Jupiter	0.5	0.4	25	116	124 (*)
Saturn	0.6	0.4	6	81	95 (*)
Uranus	0.7	0.4	1	53	59 (*)
Neptune	0.7	0.4	0.4	42	59 (*)
Pluto	0.5	0.1	0.4	38	≈ 50

(*) See footnote of Table 2.1.

$$(1-\rho)\frac{f}{4}T_s^4 - (1-\gamma)T_p^4 = 0\ .$$

This yields

$$T_p = \frac{(1-\rho)^{1/4}}{(1-\gamma)^{1/4}}\ \frac{f^{1/4}}{\sqrt{2}}\ T_s\ . \tag{2.6}$$

The number ρ is known as the albedo of the planet. The number γ has no standard name, as the above model is only a crude description of the greenhouse phenomenon. We will call γ the greenhouse coefficient. Table 2.2 gives approximate values of both ρ and γ for the nine planets. The resulting quantity $Z = (1-\rho)S$, i.e. that part of the solar constant that is actually absorbed by the planet, is also tabulated. Finally, the resulting T_p according to eqn (2.6) turns out to resemble closely the experimental T_{exp}.

It should be remarked that the cloud temperatures of Jupiter, Saturn, Uranus, and Neptune, well forecasted by Table 2.1, now exceed their theoretical value T_p of Table 2.2 systematically. What is the origin of this new discrepancy? Because the solar constant of these distant planets is so weak, we have to consider wether secondary light sources are responsible for supplementary heating. Secondary sources of light are, for example, the stars and the moons. In the next three sections we will calculate their influence on the thermal balance of the planets.

Because we are not interested in corrections of corrections, we will calculate the ΔT_p caused by stars and moons, without taking into consideration albedos and greenhouse effects. Therefore in the next three sections, calculations are performed with $\rho = 0$ and $\gamma = 0$.

2.3 Starlight ♡

A planet is not only illuminated by the Sun, but also by the other stars. Each star has its own temperature $T_\star$ and its own dilution factor $f_\star$. Most values of $T_\star$ are in the range 3000 K to 18 000 K. The values of $f_\star$ range from zero (all stars that are too distant to be perceptible) to about 10^{-20} (the nearest stars).

The thermal balance of the planet, taking into account the twinkling of the stars, is now

$$4\pi R^2\sigma \left[\frac{f}{4}T_s^4 + \Sigma\frac{f_\star}{4}T_\star^4 - (T_p + \Delta T_p)^4 \right] = 0 \ .$$

With the simple Taylor approximation

$$(T_p + \Delta T_p)^4 \approx T_p^4 + 4T_p^3\Delta T_p \ ,$$

this leads to

$$\Delta T_p \approx \frac{1}{4} \left(\frac{\Sigma f_\star T_\star^4}{f T_s^4}\right) T_p \ .$$

In our solar system $\sigma\Sigma f_\star T_\star^4$ amounts to about 20 μW/m^2, i.e. much weaker than the solar constant $\sigma f T_s^4$ of any planet. The factor $(\Sigma f_\star T_\star^4 / f T_s^4)$ ranges from about 10^{-9} for Mercury to about 10^{-5} for Pluto. Therefore the relative influence of starlight on the planet temperature is limited to the approximate range

$$10^{-10} \leq \frac{\Delta T_p}{T_p} \leq 10^{-6} \ .$$

2.4 Moonlight ♡

A planet is not only illuminated by the direct light from the Sun, but also by the light of its own moons. Let us therefore calculate the influence of the moonlight on the thermal balance of the planet.

Suppose a particular planet has a particular moon. Because the distance Sun–moon is, within very good accuracy, equal to the distance Sun–planet, the moon is irradiated by the same solar constant as the planet: $\sigma f T_s^4$.

Because it is in thermal equilibrium, it radiates an energy density $Q_\oslash$, given by

$$4\pi R_\oslash^2 (\ \sigma \frac{f}{4} T_s^4 - Q_\oslash\ ) = 0\ ,$$

where $R_\oslash$ denotes the radius of the moon. Hence

$$Q_\oslash = \frac{\sigma f T_s^4}{4}\ , \tag{2.7}$$

where $Q_\oslash$ consists partly of reflected sunlight (if the moon's albedo is non-zero) and partly of infrared thermal radiation.

Although in reality a moon has a bright and a dark side, we assume, for sake of simplicity, that the radiation $Q_\oslash$ is emitted isotropically. This radiation reaches the planet with a dilution factor

$$f_\oslash = (\frac{R_\oslash}{r_\oslash})^2\ ,$$

where $r_\oslash$ is the distance moon–planet.

The thermal balance of the planet, taking into account the moonlight, is now

$$4\pi R^2 [\ \sigma \frac{f}{4} T_s^4 + \Sigma \frac{f_\oslash}{4} Q_\oslash - \sigma (T_p + \Delta T_p)^4\ ] = 0\ .$$

With eqn (2.7), this becomes

$$\Delta T_p \approx \frac{\Sigma f_\oslash}{16}\ T_p\ .$$

As $\Sigma f_\oslash$ is a small number, we see that the heating of a planet by its moons is very weak. The factor $\Sigma f_\oslash$ ranges from zero for Mercury and Venus (which have no moons) to 3.3×10^{-5} for Jupiter (with its moons Io, Ganymede, Europa, Callisto and a series of lesser moons, which scarcely contribute to $\Sigma f_\oslash$). Therefore, the relative influence of moonlight on the planet temperature is limited to the range

$$0 \leq \frac{\Delta T_p}{T_p} \leq 2.1\ \times\ 10^{-6}\ .$$

2.5 Radiation from 'empty' space ♡

Apart from sunlight, moonlight and starlight, the planets are still illuminated by a fourth source of black-body radiation: the microwave radiation from 'empty space'. This radiation is a remnant of the Big Bang. The universe started from a dense and hot primeval fireball, that since the Big Bang expanded to its present scale. As it expanded, radiation cooled down to its present temperature T_0, i.e. 2.74 K. Because of Wien's displacement law, the associated black-body radiation consists of low-energy photons. Such radiation is called microwave radiation. The cosmic background radiation was discovered in 1965 by Penzias and Wilson, with the help of microwave antennas.

Figure 2.2 shows the spectrum $En(E)$ of the cosmic radiation. A 2.74 K Planck spectrum fits the data very well. The corresponding dilution factor f equals unity.

The thermal balance of the planet, taking into account the cosmic background, is now

$$4\pi R^2 [\ \frac{f}{4} T_s^4 + \frac{1}{4} T_0^4 - (T_p + \Delta T_p)^4\ ] = 0$$

or

$$\Delta T_p \approx \frac{1}{16} (\frac{T_0}{T_p})^4\ T_p\ .$$

The factor T_0^4/T_p^4 ranges from 1.3×10^{-9} for Mercury to 1.3×10^{-5} for Pluto. Therefore, the relative influence of the cosmic radiation on the planet temperature is limited to the range

$$8.1 \times 10^{-11} \leq \frac{\Delta T_p}{T_p} \leq 8.1 \times 10^{-7}\ .$$

2.6 Internal heat sources ♡

From the previous sections, we have to conclude that neither moonlight, nor starlight, nor cosmic radiation is responsible for the systematic error ΔT_p of the temperature of the four giant planets Jupiter, Saturn, Uranus and Neptune.

The discrepancy is explained by the presence of internal heat sources. The internal heat represents the conversion of the gravitational potential energy that became available as each planet contracted after its creation. The interior of the giant planets is still hot today. This accumulated heat leaks out very slowly, through the surface, contributing to the surface temperature in a non-negligible way.

For the other five planets, the leakage of interior heat, per unit of surface area, is very small compared with the solar constant and can therefore be neglected in the thermal balance.

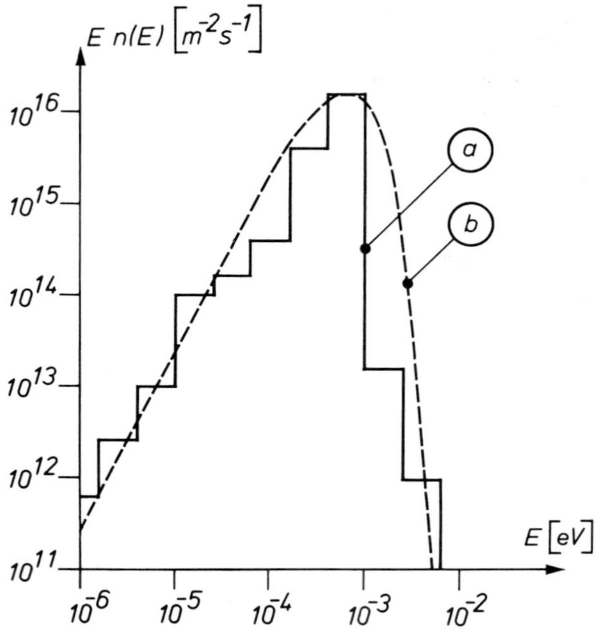

Fig. 2.2. The cosmic spectrum with logarithmic scales: (a) experimental spectrum, (b) Planck spectrum.

2.7 Conclusion

In this chapter we have encountered various black-body radiations, such as:

- the low-temperature and therefore low-energy (or microwave) radiation of the cosmic background: $T = 2.74$ K,
- the intermediate-temperature and therefore intermediate-energy (or far-infrared) radiation of the planets and moons: e.g. $T = 288$ K, and
- the high-temperature and therefore high-energy (or near-infrared and visible) radiations of the stars and Sun: e.g. $T = 5762$ K.

This explains the choices made in Fig. 1.7 of the previous chapter.

We demonstrated that cosmic radiation, starlight and moonlight can be neglected for the thermal balance of any of the planets of our solar system. Only the following quantities have an influence:

- the incident solar influx or solar 'constant' S,
- the planet's albedo ρ,
- the greenhouse effect of the planet's atmosphere, and
- the planet's internal heat source.

In Chapter 4, we will take into account only the first three effects. Thus we will use the formula

$$T_p = \frac{(1-\rho)^{1/4}}{(1-\gamma)^{1/4}} \frac{f^{1/4}}{\sqrt{2}} T_s \ .$$

Neglecting the geothermal energy source will lead to small errors for the planets Jupiter, Saturn, Uranus, and Neptune.

In Chapters 5 to 10, we will simplify reality even more, by also disregarding albedo and greenhouse effects. Thus we will use the simple formula

$$T_p = \frac{f^{1/4}}{\sqrt{2}} T_s \ .$$

This supplementary simplification will lead to considerable errors for the surface of the planet Venus, but only small errors for all the other planets.

We conclude this chapter by remarking that in all the above sections we have calculated thermal exchanges. Different bodies at different temperatures are in thermal balance. Heat is only exchanged: no heat is converted into work.

For the conversion of solar energy into work, we need to have a theory of the conversion of heat into work. In Chapter 3, we will extensively develop such a theory: thermodynamics. In Chapters 4 to 10, we will apply this thermodynamics to the conversion of the solar radiation.

Bibliography

1. M. Thekaekara (1973). Solar energy outside the Earth's atmosphere. *Solar Energy*, volume 14, pp. 109 - 127.

2. C. Backus (1976). Introduction. In *Solar cells* (ed. C. Backus). I.E.E.E. Press, New York, pp. 1 - 9.

3. W. Palz (1978). *Solar electricity*. Butterworths-Unesco, Paris, pp. 35 - 52.

4. J. Beatty, B. O'Leary and A. Chaikin (1981). *The new solar system*. Cambridge University Press, Cambridge.

5. T. Encrenaz and J. Bilring (1987). *Le système solaire.* Presse C.N.R.S., Paris.

6. M. Waldrop (1990). Cosmic Background Explorer confronts cosmic conundrums. *Science*, volume 247, pp. 411 - 413.

3

Thermodynamical engines

3.1 Introduction

In this chapter we will construct a thermodynamic framework for the description of the conversion of solar energy into work. This will be done by introducing successively three thermodynamic models:

- the Carnot engine,
- the Curzon–Ahlborn engine, and
- the Stefan–Boltzmann engine.

We will develop these models rather axiomatically: by accepting two axioms:

- conservation of energy, and
- conservation of entropy,

the former being valid anywhere, the latter being valid only in reversible parts of the engines.

These axioms are supplemented by the laws governing heat transport. These heat conduction laws are progressively more complicated for the three models:

- the Carnot engine contains no heat conductors,
- the Curzon–Ahlborn engine contains linear conductors, and
- the Stefan–Boltzmann engine contains non-linear conductors.

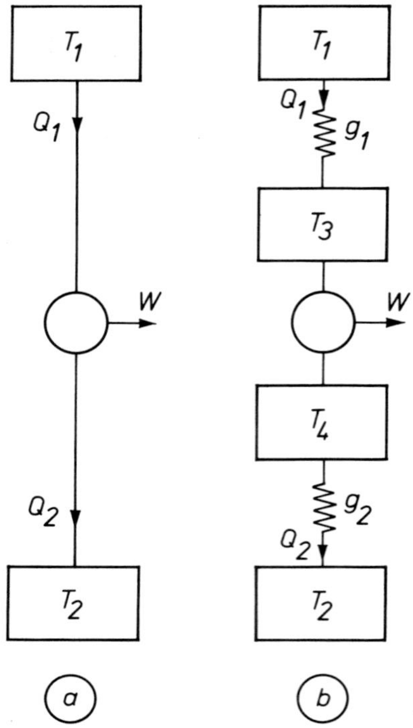

Fig. 3.1. Two thermodynamic configurations: (a) Carnot engine, (b) Curzon–Ahlborn engine.

3.2 The Carnot engine

Figure 3.1(a) shows the basic features of the Carnot theory:

- two heat reservoirs: one at the 'high' temperature T_1 and one at the 'low' temperature T_2;
- a reversible Carnot engine between these two reservoirs.

The words 'high' and 'low' stand between quotes, because we will sometimes consider the case where $T_1 < T_2$. The adjectives are used because normally $T_1 > T_2$.

In order to determine the properties of the structure, we only have to implement the two laws of thermodynamics:

Axiom 1: conservation of energy:

$$\Sigma Q + \Sigma W = 0 \ ,$$

where Q are the heat flows per unit of time, W the work flows per unit of time, and Σ denotes a summation over a closed surface around the engine.

Axiom 2: conservation of entropy:

$$\Sigma S = 0 \ ,$$

where S are the entropy flows per unit of time.

If we postulate that any heat flow Q is accompanied by an entropy flow Q/T and any work flow W is accompanied by a zero entropy flow, Axiom 2 becomes

$$\Sigma \frac{Q}{T} = 0 \ .$$

Because of Axiom 1 we have

$$Q_1 = W + Q_2 \ . \tag{3.1}$$

Because of Axiom 2 we have

$$Q_1/T_1 = Q_2/T_2 \ . \tag{3.2}$$

If we introduce the following definition of the conversion efficiency η:

$$W = \eta Q_1 \ ,$$

then we obtain the following from eqns (3.1) and (3.2)

$$\eta = 1 - T_2/T_1 \ , \tag{3.3}$$

which is nothing other than the classical proof of Carnot's formula, for a reversible engine working between the temperatures T_1 and T_2. If we call the efficiency of a Carnot engine the Carnot efficiency and denote it η_C, we thus have

$$\eta_C = 1 - \frac{T_2}{T_1} \ .$$

We have to distinguish two modes of operation of the Carnot engine:

- If $W > 0$, then the engine delivers a positive work flow; it is said to function as a heat engine.

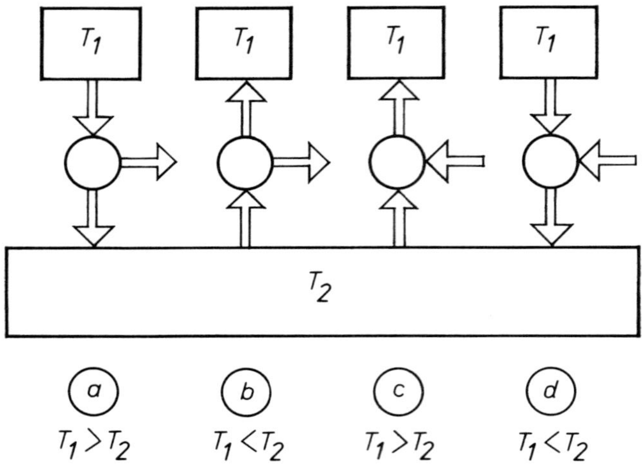

Fig. 3.2. Four thermodynamic modes: (a) heat engine, (b) heat engine, (c) heat pump, (d) refrigerator. Note: all arrows indicate positive energy flows.

- If $W < 0$, then the engine delivers a negative work flow. In other words, we have to supply a positive work flow $-W$ to the engine. We now distinguish two subcases:

  * If the 'surroundings' at temperature T_2 is the cooler reservoir (i.e. $T_1 > T_2$), then the engine is said to function as a heat pump: $\eta > 0$ and thus $Q_1 < 0$. The work flow we supply is used to heat reservoir 1, by maintaining a heat flow from reservoir 2 to reservoir 1, in spite of the natural tendency of heat to flow from hot to cold.

  * If the 'surroundings' at temperature T_2 is the hotter reservoir (i.e. $T_1 < T_2$), then the engine is said to function as a refrigerator: $\eta < 0$ and thus $Q_1 > 0$. The work flow we supply is used to cool reservoir 1, by maintaining a heat flow from reservoir 1 to reservoir 2, in spite of the natural tendency of heat to flow from hot to cold.

See Fig. 3.2 .

Readers familiar with the Carnot theory have probably recognized the classical scheme, as it appears in most physics textbooks. It should be stressed, however, that the usual description of the Carnot converter is in terms of heat Q, work W, and entropy S (in J, J, and J/K, respectively), whereas the above description is in terms of heat flow Q, work flow W, and entropy flow S (in W, W, and W/K, respectively). These quantities are thus per unit of time. This will remain so for the rest of this book, even when this is not indicated at every occasion. We investigate steady-state situations and all extensive variables are time-independent transports from one place (i.e. reservoir) to another and are thus calculated per unit of time.

3.3 The Curzon–Ahlborn engine

Figure 3.1(b) shows the different features of the Curzon–Ahlborn engine:

- two heat reservoirs: one at the 'high' temperature T_1 and one at the 'low' temperature T_2;
- two irreversible components: thermal resistors which limit the heat flows Q_1 and Q_2; they cause a temperature drop from T_1 to an intermediate temperature T_3 and a temperature drop from an intermediate temperature T_4 to T_2;
- a reversible Carnot engine between the intermediate heat reservoir at temperature T_3 and the intermediate reservoir at temperature T_4.

In order to determine the properties of the new structure, we have to implement again the two laws of thermodynamics:

Axiom 1: conservation of energy:

$$\Sigma Q + \Sigma W = 0 \ ,$$

where Q are the heat flows, W the work flows, and Σ denotes a summation over an arbitrary closed surface.

Axiom 2: conservation of entropy:

$$\Sigma \frac{Q}{T} = 0 \ ,$$

where Σ denotes a summation over a closed surface containing only reversible parts, i.e. a surface enclosing only the Carnot part.

Because of Axiom 1 we have

$$Q_1 = W + Q_2 \ . \tag{3.4}$$

Because of Axiom 2 we have

$$Q_1/T_3 = Q_2/T_4 \ . \tag{3.5}$$

If we again introduce the conversion efficiency η:

$$W = \eta Q_1 \ ,$$

we then obtain from eqns (3.4) and (3.5)

$$\eta = 1 - T_4/T_3 \ , \tag{3.6}$$

which is none other than Carnot's formula, however for a reversible engine working between the temperatures T_3 and T_4 (instead of T_1 and T_2, as in Section 3.2).

We now need constitutive laws for the thermal conductors. In accordance with Curzon and Ahlborn [1], we use linear laws:

$$Q_1 = g_1(T_1 - T_3) \tag{3.7}$$

$$Q_2 = g_2(T_4 - T_2) \ , \tag{3.8}$$

after Fourier's law for the diffusion of heat:

$$Q = -\lambda \nabla T \ .$$

The proportionality coefficients g_1 and g_2 are called the thermal conductances or heat conductances.

Substitution of eqns (3.7) and (3.8) into eqn (3.5) yields

$$\frac{g_1(T_1 - T_3)}{T_3} = \frac{g_2(T_4 - T_2)}{T_4} \ . \tag{3.9}$$

Solving the set of two eqns (3.6) and (3.9) for T_3 and T_4 gives

$$T_3 = \frac{g_1}{g_1 + g_2} T_1 + \frac{g_2}{g_1 + g_2} \frac{1}{1 - \eta} T_2$$

$$T_4 = \frac{g_1}{g_1 + g_2} (1 - \eta) T_1 + \frac{g_2}{g_1 + g_2} T_2 \ .$$

Substitution of the T_3-result into eqn (3.7) finally yields

$$Q_1 = g \frac{T_1 - T_2 - T_1\eta}{1 - \eta} \ , \tag{3.10}$$

where

$$g = \frac{g_1 g_2}{g_1 + g_2} \ .$$

This relationship between Q_1 and η, which we will call the 'current-efficiency characteristic', is depicted in Fig. 3.3(a). In the figure we have also displayed the two asymptotes, as well as two particular points on the $Q_1(\eta)$ characteristic:

- For $\eta = -\infty$, the heat current Q_1 equals its saturation value:

$$Q_{sat} = Q_1(-\infty) = gT_1 \ . \tag{3.11}$$

- The intersection with the ordinate axis is called the short-circuit point. The short-circuit current equals

$$Q_{sc} = Q_1(0) = g(T_1 - T_2) \ .$$

- The intersection with the abscissa axis is called the open-circuit point. By equating the right-hand side of eqn (3.10) to zero and solving for η, we find the open-circuit efficiency:

$$\eta_{oc} = 1 - \frac{T_2}{T_1} \ ,$$

  i.e. exactly the Carnot efficiency η_C of a reversible engine working between the temperatures T_1 and T_2.

- For $\eta = 1$, the heat current becomes infinite:

$$Q_1(1) = -\infty \ .$$

Multiplication of eqn (3.10) by η gives

$$W = g \ \frac{\eta(T_1 - T_2 - T_1\eta)}{1 - \eta} \ . \tag{3.12}$$

We call $W(\eta)$ the 'work-efficiency' characteristic. See Fig. 3.3(b).

The short-circuit point and the open-circuit point divide the $Q_1(\eta)$ and $W(\eta)$ characteristics into three parts:

- For $-\infty < \eta < 0$, we have

$$\begin{aligned} T_1 &> T_3 \\ T_4 &> T_3 \\ T_4 &> T_2 \ , \end{aligned}$$

  and the engine between T_3 and T_4 works as a refrigerator.

- For $0 < \eta < \eta_{oc}$, we have

$$T_1 > T_3 > T_4 > T_2 \ ,$$

  and the engine between T_3 and T_4 works as a heat engine.

- For $\eta_{oc} < \eta < 1$, we have

$$T_3 > T_1 > T_2 > T_4 \ ,$$

  and the engine between T_3 and T_4 works as a heat pump.

To demonstrate that the Curzon–Ahlborn engine is not reversible, we calculate the ΣS for a closed surface containing the whole engine, i.e. both the Carnot engine and the two heat conductors. We have

$$\Sigma S = \frac{-Q_1}{T_1} + \frac{Q_2}{T_2} \ ,$$

where the minus sign of Q_1 takes into account the fact that the heat Q_1 enters the surface Σ instead of leaving it. Thus

$$\begin{aligned} \Sigma S &= Q_2/T_2 - Q_1/T_1 \\ &= (Q_1 - W)/T_2 - Q_1/T_1 \\ &= \frac{Q_1}{T_2} - \frac{\eta Q_1}{T_2} - \frac{Q_1}{T_1} \\ &= (\eta_C - \eta)\frac{Q_1}{T_2} \ , \end{aligned} \tag{3.13}$$

where η_C denotes the Carnot efficiency $1 - T_2/T_1$. Substitution of eqn (3.10) into eqn (3.13) yields

$$\Sigma S = g \ \frac{(T_1 - T_2 - T_1\eta)^2}{T_1 T_2 (1 - \eta)} \ .$$

Figure 3.3 shows the quantity $\Sigma S(\eta)$, together with the $Q_1(\eta)$ and $W(\eta)$ characteristics.

We see that under reversible conditions, i.e. $\eta = 1 - T_4/T_3$ equal to its Carnot value $\eta_C = 1 - T_2/T_1$, both Q_1 and W are zero. This expresses the general principle that under reversible conditions conversion happens infinitely slowly. We see further that, for this same value of η, the entropy creation rate ΣS is both minimum and zero. This expresses the fact that in a reversible system entropy is constant, whereas in an irreversible system entropy is increasing.

We now remark that the produced work W displays a maximum at an η value between 0 and η_C. This value is found by applying $dW/d\eta = 0$ to eqn (3.12), yielding immediately

$$T_1\eta^2 - 2T_1\eta + T_1 - T_2 = 0 \ .$$

Solving for η gives

$$\eta = 1 - \sqrt{\frac{T_2}{T_1}} \ . \tag{3.14}$$

This important formula was discovered by Curzon and Ahlborn [1] in 1975. Under this maximum-power condition, the quantities Q_1, W and ΣS take non-zero values:

$$\begin{aligned} Q_1 &= g\,\sqrt{T_1}(\sqrt{T_1} - \sqrt{T_2}) \\ W &= g\,(\sqrt{T_1} - \sqrt{T_2})^2 \\ \Sigma S &= g\,(\sqrt{T_1} - \sqrt{T_2})^2/\sqrt{T_1T_2} \ . \end{aligned}$$

Thus we have derived the Curzon–Ahlborn formula, which expresses the efficiency of the engine in maximum-power condition. Figure 3.4 shows the relation between η and the temperature ratio T_2/T_1. We see that the Curzon–Ahlborn curve lies substantially below the straight line $\eta = 1 - T_2/T_1$, which expresses the Carnot efficiency η_C for the same engine, but working in reversible conditions. This confirms the general rule: efficiency under irreversible conditions is lower than under reversible conditions.

The Curzon–Ahlborn efficiency $\eta_{C\&A}$ not only holds for linear thermal conductors in combination with a Carnot engine, but it turns out to be also a very good approximation for linear conductors combined with other cyclic engines, such as Otto, Joule and Diesel engines [6] [7] [8].

3.4 Endoreversible engines

In 1979, Rubin [2] introduced the following definition: " Endoreversible engines are irreversible engines where all irreversibilities are restricted to the coupling of the engine to the external world. ". It is clear that the Curzon–Ahlborn engine, described in Section 3.3, is such an engine.

The heart or inner part ($\epsilon\nu\delta o$ or *endo* is Greek for internal) of the Curzon–Ahlborn engine is indeed a reversible Carnot engine. All irreversibilities and thus all the entropy generation ΣS take place in the coupling of the latter engine to the external world, in other words, in the heat exchange between the Carnot engine and the two external heat reservoirs, i.e. in the heat conductors g_1 and g_2. Irreversibilities caused by turbulences, by mixing, by friction, and so on, all are assumed to be absent.

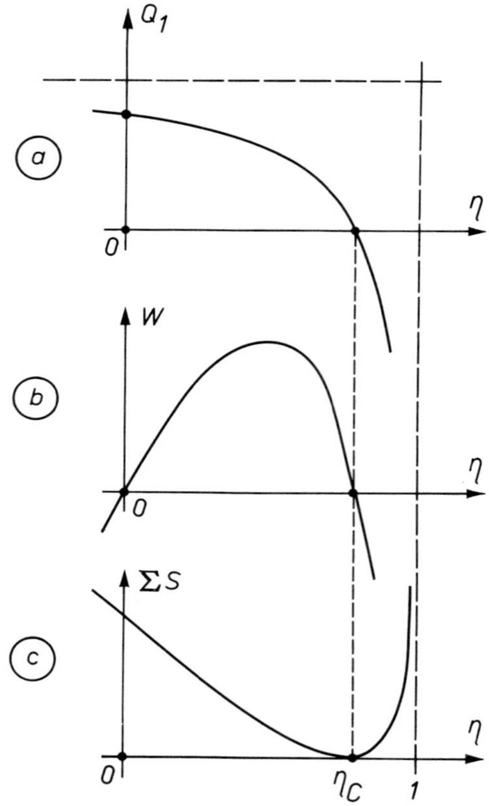

Fig. 3.3. Thermodynamic quantities as a function of the conversion efficiency $\eta = 1 - T_4/T_3$ of a Curzon–Ahlborn engine: (a) heat consumption Q_1, (b) work production W, (c) entropy creation rate ΣS.

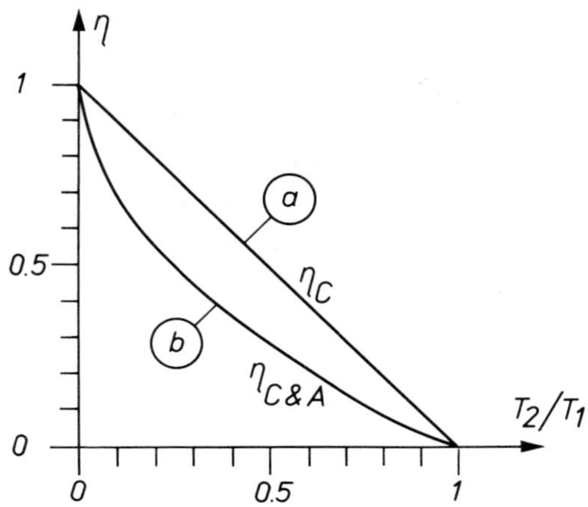

Fig. 3.4. Energy conversion efficiency η as a function of temperature ratio T_2/T_1: (a) according to Carnot, (b) according to Curzon and Ahlborn.

It is interesting to note that the Curzon–Ahlborn engine is an endoreversible engine *avant la lettre*, as it was published in 1975, four years before Rubin introduced his definition of endoreversible engines.

3.5 An example

In order to illustrate the formula of Curzon and Ahlborn, let us look at an example [4] [12]: the nuclear power plant 'Doel 4' at Doel (Belgium). It is a rather conventional plant with a pressurized water reactor. See Figs. 3.5 and 3.6.

In the secondary heat circuit are located the turbine and the alternator, which together convert the heat Q_1 into the electrical power W. The heat reservoir at high temperature is the primary loop which has an average temperature of $T_1 = 293$ °C. The steam generator acts as thermal conductor g_1. The thermal conductor g_2 consists of the condensor, the tertiary circuit, the cooling tower, and the heat exchanger with the river Schelde. The heat reservoir at low temperature is composed of the atmosphere and the river. Its temperature varies a little with the seasons, but we can say that on average $T_2 \approx 10$ °C. If we consider, against all odds, that the nuclear plant is a reversible engine, then Carnot's formula predicts

$$\eta = 1 - \frac{T_2}{T_1} = 1 - \frac{283\ \text{K}}{566\ \text{K}} = 0.500\ .$$

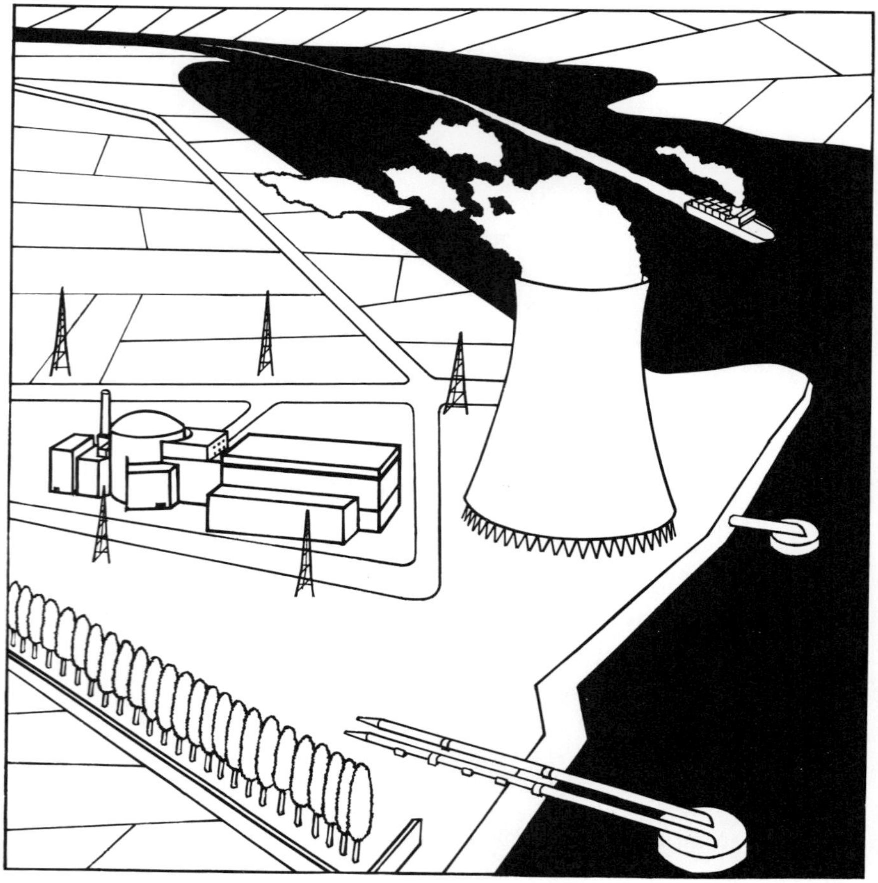

Fig. 3.5. Nuclear power plant 'Doel 4'.

If we assume that the plant is an endoreversible engine working in maximum-power point, then the Curzon–Ahlborn formula predicts

$$\eta = 1 - \sqrt{\frac{T_2}{T_1}} = 1 - \sqrt{\frac{283 \text{ K}}{566 \text{ K}}} = 0.293 \ .$$

The experimental value of Q_1 is 3000 MW and the experimental value of W equals 1049 MW, so that

$$\eta = \frac{1049 \text{ MW}}{3000 \text{ MW}} = 0.350 \ .$$

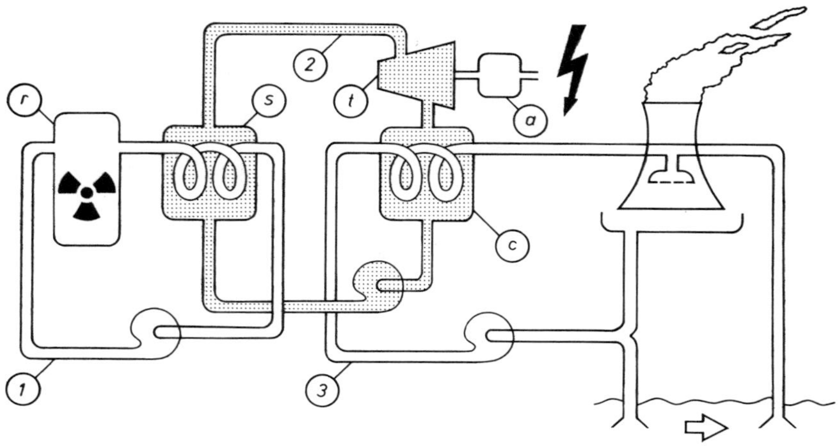

Fig. 3.6. Nuclear power plant: (1) primary circuit, (2) secondary circuit, (3) tertiary circuit, (r) reactor, (s) steam generator, (t) turbine, (a) alternator, (c) condensor.

We see that the Curzon–Ahlborn value of 29 per cent is closer to the experimental value of 35 per cent than the Carnot value of 50 per cent is. But why is the actual efficiency η larger than the value $\eta_{C\&A}$? The explanation comes from economical considerations.

Let us consider Fig. 3.7, which roughly recalls Fig. 3.3(b). We recognize a maximal W at η equal to the Curzon–Ahlborn value $1-\sqrt{T_2/T_1}$ and a W equal to zero at η equal to the Carnot value $1-T_2/T_1$. From an economical point of view neither of these two points is the optimal working point. Indeed:

- The higher the delivered power W, the faster the investments for building the plant will be recuperated.
- The higher the efficiency η, the better the expenses for the procurement of primary energy (*in casu* uranium) will be recuperated.

When we make η somewhat larger than $\eta_{C\&A}$, then the return of the fission material procurement will grow, without a noteworthy lowering of the return of the infrastructure investment. Indeed, at the Curzon–Ahlborn point we have $dW/d\eta = 0$, such that the loss dW can be neglected, even for a non-negligible gain $d\eta$. The exact location of the economically optimal working point on the $W(\eta)$ characteristic depends on the actual ratio of fixed costs to variable costs. In any case we have

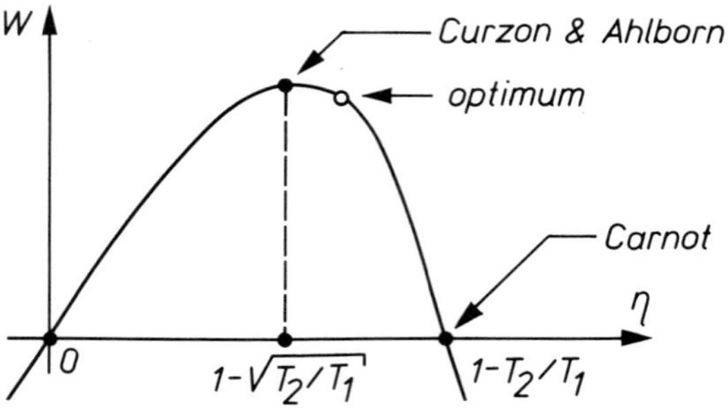

Fig. 3.7. Delivered power W as a function of efficiency $\eta = 1 - T_4/T_3$.

$$\eta_{C\&A} < \eta_{opt} < \eta_C \ .$$

From the above it is clear that the duality ' reversible efficiency vs. maximum-power efficiency ' is related to the duality ' free investments vs. free fuel '. It is therefore no surprise that endoreversible thermodynamics is especially suited to describe the conversion efficiency of renewable energy sources (i.e. of energy sources with free fuel) in general, and solar energy in particular.

However, the Curzon–Ahlborn model is not immediately applicable to solar energy, as the transport of solar heat does not satisfy the linear law

$$Q_1 = g_1(T_1 - T_3) \ ,$$

but instead a Stefan–Boltzmann law:

$$Q_1 = g_1(T_1^4 - T_3^4) \ ,$$

as we have seen in Sections 1.5 - 1.7. Therefore, it is necessary to transform the Curzon–Ahlborn engine into an endoreversible engine where transports obey the Stefan–Boltzmann law. We will call such engines Stefan–Boltzmann engines.

3.6 The Stefan–Boltzmann engine

If heat is delivered to and drained from the Carnot engine by means of radiation instead of by conduction, then the transport eqns (3.7)-(3.8) have to be replaced by the corresponding Stefan–Boltzmann laws:

$$Q_1 = g_1(T_1^4 - T_3^4) \qquad (3.15)$$

$$Q_2 = g_2(T_4^4 - T_2^4) \ . \qquad (3.16)$$

Proceeding exactly as in Section 3.3, i.e. by solving the set of equations

$$\frac{g_1(T_1^4 - T_3^4)}{T_3} = \frac{g_2(T_4^4 - T_2^4)}{T_4}$$
$$\eta = 1 - \frac{T_4}{T_3}$$

for T_3 and T_4, now gives

$$T_3^4 = \frac{g_1}{g_1 + g_2(1-\eta)^3}\, T_1^4 + \frac{g_2}{g_1 + g_2(1-\eta)^3}\, \frac{1}{1-\eta}\, T_2^4$$
$$T_4^4 = \frac{g_1}{g_1 + g_2(1-\eta)^3}\, (1-\eta)^4 T_1^4 + \frac{g_2}{g_1 + g_2(1-\eta)^3}\, (1-\eta)^3 T_2^4 \, .$$

Substitution of the T_3-result into eqn (3.15) yields

$$Q_1 = g_1 g_2 \, \frac{(1-\eta)^4 T_1^4 - T_2^4}{g_1(1-\eta) + g_2(1-\eta)^4} \, . \tag{3.17}$$

This relationship between Q_1 and η, which we will again call the current-efficiency characteristic, is depicted in Fig. 3.8(a). In this figure we have again displayed the two asymptotes, as well as the two particular points on the $Q_1(\eta)$ characteristic:

- For $\eta = -\infty$, the heat current Q_1 equals its saturation value:

$$Q_{sat} = Q_1(-\infty) = g_1 T_1^4 \, . \tag{3.18}$$

- For $\eta = 0$, we find the short-circuit current:

$$Q_{sc} = Q_1(0) = g(T_1^4 - T_2^4) \, ,$$

  where

$$g = \frac{g_1 g_2}{g_1 + g_2} \, .$$

- By solving $Q_1(\eta) = 0$ for η, we find the open-circuit efficiency:

$$\eta_{oc} = 1 - \frac{T_2}{T_1} \, ,$$

  i.e. again exactly the Carnot efficiency η_C.

- For $\eta = 1$, the heat current becomes infinite:

$$Q_1(1) = -\infty \, .$$

As in the Curzon–Ahlborn case, the short-circuit point and the open-circuit point devide $Q_1(\eta)$ into three parts:

- For $-\infty < \eta < 0$, the engine between T_3 and T_4 works as a refrigerator.
- For $0 < \eta < \eta_{oc}$, the engine between T_3 and T_4 works as a heat engine.
- For $\eta_{oc} < \eta < 1$, the engine between T_3 and T_4 works as a heat pump.

The curves of Fig. 3.8 are very similar to the corresponding curves on Fig. 3.3 for the Curzon–Ahlborn model. But we see that the curve $Q_1(\eta)$ is more 'rectangular' in the Stefan–Boltzmann case than in the Curzon–Ahlborn case. Whereas the curve $Q_1(\eta)$ is a simple hyperbola in the Curzon–Ahlborn case, it is a fifth-degree curve in the present case. It is therefore no surprise that maximum-power values of η and Q_1 cannot be expressed in closed form.

Multiplying $Q_1(\eta)$ by η yields $W(\eta)$, the work-efficiency characteristic:

$$W(\eta) = g_1 g_2 \, \eta \, \frac{(1-\eta)^4 T_1^4 - T_2^4}{g_1(1-\eta) + g_2(1-\eta)^4} \; .$$

It is displayed in Fig. 3.8(b). The curve displays a maximum for $dW/d\eta = 0$, i.e. for the η value satisfying

$$\begin{aligned} g_2 T_1^4 (1-\eta)^8 \; &+ \; 4 g_1 T_1^4 (1-\eta)^5 \; + \; 3(g_2 T_2^4 - g_1 T_1^4)(1-\eta)^4 \\ &- \; 4 g_2 T_2^4 (1-\eta)^3 \; - \; g_1 T_2^4 = 0 \; . \end{aligned} \qquad (3.19)$$

The solution η of this eighth-degree equation should yield a Stefan-Boltzmann equivalent of the Curzon–Ahlborn formula. But we can immediately make two painful conclusions:

- As an eighth-degree equation has no analytical solution, there will be no closed form analogous to eqn (3.14).
- In contrast to the Curzon–Ahlborn case, the solution for the Stefan–Boltzmann case will be dependent of the conductance ratio g_1/g_2.

Thus the Stefan–Boltzmann engine leads to numerical calculations. Figure 3.9 shows the computed value of the maximum-power efficiency η, for three different cases:

- $g_1 \ll g_2$, i.e. $g_1/g_2 \to 0$,
- $g_1 = g_2$, i.e. $g_1/g_2 = 1$, and

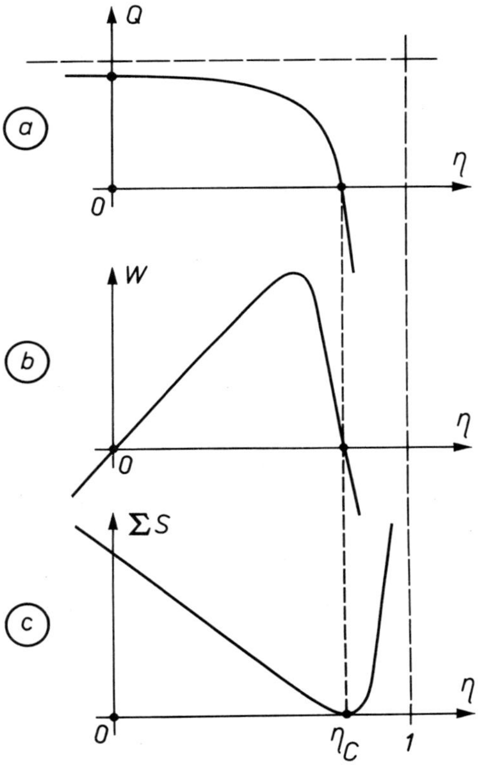

Fig. 3.8. Thermodynamic quantities as a function of the conversion efficiency $\eta = 1 - T_4/T_3$ of a Stefan–Boltzmann engine: (a) heat consumption Q_1, (b) work production W, (c) entropy creation rate ΣS.

- $g_1 \gg g_2$, i.e. $g_1/g_2 \to \infty$.

The general Stefan–Boltzmann engine was introduced by De Vos [3]. The first special case was extensively treated by De Vos [3] [5] and by Chen and Yan [9], whereas Gordon [11] made a detailed comparison between the three special cases.

Note that even in the above peculiar cases the mathematics cannot be performed analytically. Indeed:

- for $g_1 = 0$, eqn (3.19) yields a fifth-degree equation:

$$T_1^4(1-\eta)^5 + 3T_2^4(1-\eta) - 4T_2^4 = 0 ,$$

- for $g_2 = 0$, it also yields a fifth-degree equation:

$$4T_1^4(1-\eta)^5 - 3T_1^4(1-\eta)^4 - T_2^4 = 0 ,$$

- whereas for $g_2 = g_1$, the equation remains of eighth degree:

$$T_1^4(1-\eta)^8 + 4T_1^4(1-\eta)^5 + 3(T_2^4 - T_1^4)(1-\eta)^4 - 4T_2^4(1-\eta)^3 - T_2^4 = 0 . \tag{3.20}$$

None of these polynomial equations has an analytical solution. Mathematicians say that there is no solution 'in radicals', i.e. no solution containing only a finite number of the operations addition, substraction, multiplication, division and extraction of root ($+$, $-$, $\times$, $/$ and $\sqrt[n]{\ }$).

We end this section by calculating the total entropy production rate ΣS. We still have

$$\begin{aligned} \Sigma S &= Q_2/T_2 - Q_1/T_1 \\ &= (1-\eta)Q_1/T_2 - Q_1/T_1 \\ &= (\eta_C - \eta)\frac{Q_1}{T_2} , \end{aligned}$$

just like in Section 3.3. Substituting the Stefan–Boltzmann characteristic $Q_1(\eta)$, i.e. eqn (3.17), we obtain

$$\Sigma S = \frac{g_1 g_2}{T_1 T_2}\,\frac{[(1-\eta)T_1 - T_2]\,[(1-\eta)^4 T_1^4 - T_2^4]}{g_1(1-\eta) + g_2(1-\eta)^4} .$$

Figure 3.8 shows $\Sigma S(\eta)$, together with the characteristics $Q_1(\eta)$ and $W(\eta)$. Note that all three quantities become zero at $\eta = \eta_C$.

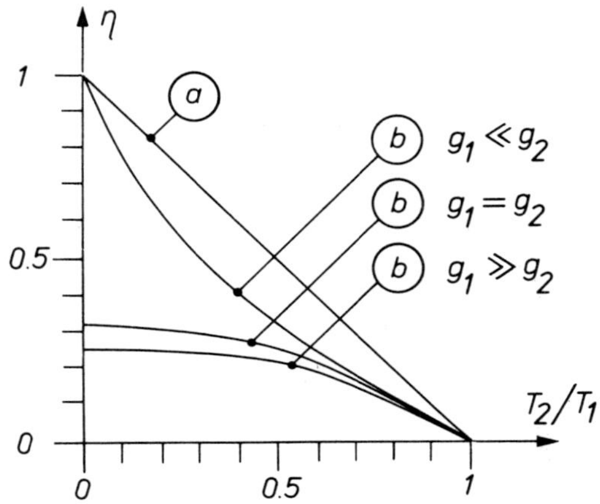

Fig. 3.9. Energy conversion efficiency η as a function of temperature ratio T_2/T_1: (a) according to Carnot, (b) according to the Stefan–Boltzmann model.

3.7 Conclusion

We recall here the main results of this chapter:

- The Carnot engine leads to the Carnot formula:

$$\eta = 1 - \frac{T_2}{T_1} \ .$$

- The Curzon–Ahlborn engine leads to the Curzon–Ahlborn formula:

$$\eta = 1 - \sqrt{\frac{T_2}{T_1}} \ .$$

- The Stefan–Boltzmann engine leads to an efficiency which cannot be written in a closed formula. The efficiency η is the solution of the polynomial equation

$$\begin{aligned} g_2 T_1^4 (1-\eta)^8 \ &+ \ 4 g_1 T_1^4 (1-\eta)^5 \ + \ 3(g_2 T_2^4 - g_1 T_1^4)(1-\eta)^4 \\ &- \ 4 g_2 T_2^4 (1-\eta)^3 \ - \ g_1 T_2^4 = 0 \ . \end{aligned}$$

The above results have following foundations:

- The Carnot efficiency is the most basic of the three, as it is based only upon the following assumptions:

  * energy is preserved everywhere,
  * entropy is preserved in reversible parts,
  * work is entropyless,
  * a heat flow Q extracted from or drained into a reservoir at temperature T is accompanied by an entropy flow Q/T.

- The Curzon–Ahlborn efficiency is based upon one more assumption, i.e. that heat transport Q between two reservoirs obeys

$$Q = g(T_1 - T_2) \ .$$

  This is only a phenomenological law, i.e. a law which cannot be deduced from first principles. Only if abs($T_1 - T_2$) is small compared to both T_1 and T_2, is the Fourier law correct.

- The fourth-power law of heat exchange by radiation can be deduced from first priciples. It was deduced in Chapter 1 from the basic laws of quantum statistics. The Stefan–Boltzmann law

$$Q = g(T_1^4 - T_2^4)$$

  is correct, even if T_1 and T_2 have different order of magnitude. Nevertheless, we can say that the efficiency of the Stefan–Boltzmann engine is somewhat less fundamental than the Carnot efficiency, as it is explicitly based upon one more assumption:

  * space is three-dimensional.

3.8 An example

We finish this chapter by describing an example of a Stefan–Boltzmann engine. The device chosen does not produce work, i.e. does not function as a heat engine. Indeed, in the following chapters we will have ample opportunity to present various Stefan–Boltzmann heat engines in detail. In this section, we will, on the contrary, discuss an example, where the Carnot part of the system works as a refrigerator: an infrared thermographic camera [10] [13].

Figure 3.10 shows such a camera, aimed at a human hand in a room. The skin temperature T_1 varies from place to place: from 20 °C on the fingertips to 30 °C on the palm of the hand. The room temperature T_2 is about 18 °C. The radiation heat detector is a so-called bolometer or pyrometer, at temperature T_3. It measures the net incoming heat flux

$$Q_1 = \sigma(T_1^4 - T_3^4) \; . \tag{3.21}$$

From the Q_1 information, it reconstructs the T_1 value and displays the latter on a computer screen, in so-called 'false colours'.

The above dependence of Q_1 upon the detector temperature T_3 is displayed in Fig. 3.11. If the detector were simply at room temperature, i.e. $T_3 = T_2$, then the signal would be

$$Q_1 = \sigma(T_1^4 - T_2^4) \; ,$$

which varies over the hand surface from 11 to 66 W/m^2. If the detector is cooled, a higher signal is at our disposal. In the limit for $T_3 \rightarrow 0$, we have that Q_1 equals σT_1^4, which varies from 418 to 473 W/m^2. Even more important than the large signal, is the fact that this signal is less sensitive to fluctuations of the detector temperature T_3. Indeed, in the limit for $T_3 \rightarrow 0$, the derivative dQ_1/dT_3 vanishes, whereas for $T_3 = T_2$ (i.e. uncooled camera), the derivative dQ_1/dT_3 equals $-\ 4\sigma T_2^3$, i.e. $-$ 1.40 W/m^{2}K. It is clear that processing the signal Q_1 is not easy, if it is so sensitive to thermal noise.

It is therefore no surprise that thermographic cameras are always cooled. This happens at the expense of power. In Fig. 3.10 the camera is cooled to $-$ 180 °C by means of the expansion of a compressed gas. The argon bottle (attached to the bottom of the camera) serves as the source of (pneumatic) power $-W$. With $T_3 = -$ 180 °C, we find a Q_1 only slightly smaller than its maximum value: from 414 to 469 W/m^2 and a derivative dQ_1/dT_3 far smaller than for the uncooled case: $-$ 0.046 W/m^{2}K.

Let us now relate the above facts with our standard description of a Stefan–Boltzmannn engine. We must therefore replace the independent variable T_3 by the dimensionless independent variable η.

For simplicity, we will assume perfect thermal contact between reservoir 4 and reservoir 2, i.e. we assume $T_4 = T_2$. Then $\eta = 1 - T_3/T_4$ equals $1 - T_3/T_2$ and eqn (3.21) becomes

$$Q_1(\eta) = \sigma \; [\; T_1^4 - \frac{T_2^4}{(1-\eta)^4} \;] \; .$$

The same result is obtained from the general expression (3.17) by postulating $g_2 = \infty$.

Figure 3.12 shows schematically the function $Q_1(\eta)$. Note that it is a remake of Fig. 3.8(a), but for a somewhat different (i.e. somewhat higher)

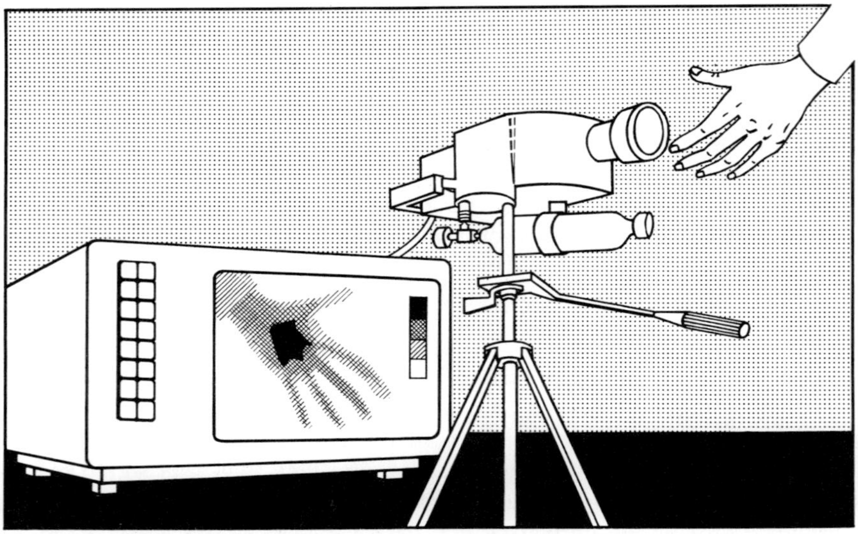

Fig. 3.10. Thermographic camera.

value of the ratio T_2/T_1. Uncooled cameras have an η equal to zero, such that the detected signal Q_1 is nothing else but the short-circuit current $Q_{sc} = Q_1(0)$. Cameras cooled to absolute zero have an η equal to $-\infty$, such that the detected signal Q_1 is the saturation current $Q_{sat} = Q_1(-\infty)$. Our camera cooled at $T_3 = -180$ °C has an η equal to -2.13. The small slope of the curve $Q_1(\eta)$ at $\eta = -2.13$ illustrates again the good noise-immunity of the detector.

Bibliography

1. F. Curzon and B. Ahlborn (1975). Efficiency of a Carnot engine at maximum power output. *American Journal of Physics*, volume 43, pp. 22 - 24.

2. M. Rubin (1979). Optimal configurations of a class of irreversible engines. *Physical Review A*, volume 19, pp. 1272 - 1276.

3. A. De Vos (1985). Efficiency of some heat engines at maximum-power conditions. *American Journal of Physics*, volume 53, pp. 570 - 573.

4. J. Van hove (1985). *Elektriciteit uit kernenergie: kerncentrale Doel.* Ebes n.v., Antwerpen.

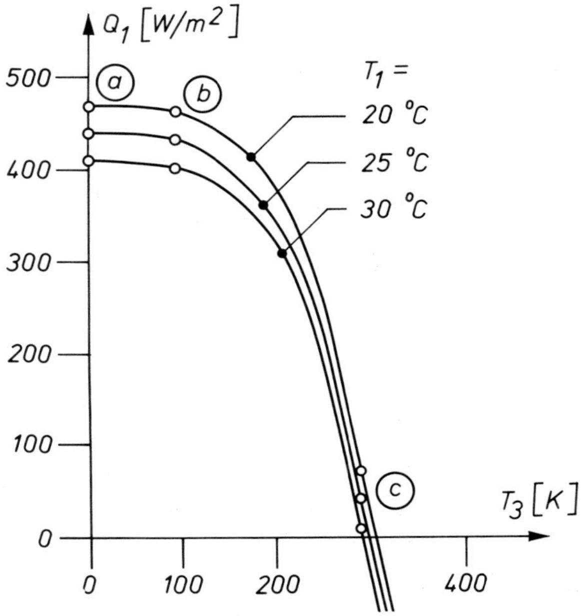

Fig. 3.11. Detected heat current as a function of temperature T_3: (a) with detector cooled to − 273 °C, (b) with detector cooled to − 180 °C, (c) with uncooled detector.

5. A. De Vos (1987). Reflections on the power delivered by endoreversible engines. *Journal of Physics D: Applied Physics*, volume 20, pp. 232 - 236.

6. H. Leff (1987). Thermal efficiencies at maximum work output: new results for old heat engines. *American Journal of Physics*, volume 55, pp. 602 - 610.

7. P. T. Landsberg and H. Leff (1989). Thermodynamic cycles with nearly universal maximum-power efficiencies. *Journal of Physics A: Mathematical and General*, volume 22, pp. 4019 - 4026.

8. J. Gordon (1989). Maximum power point characteristics of heat engines as a general thermodynamic problem. *American Journal of Physics*, volume 57, pp. 1136 - 1142.

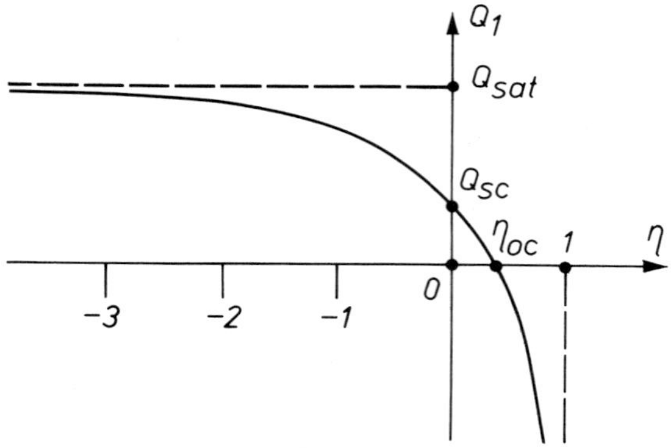

Fig. 3.12. Detected heat current as a function of engine efficiency η.

9. L. Chen and Z. Yan (1989). The effect of heat-transfer law on performance of a two-heat-source endoreversible cycle. *Journal of Chemical Physics*, volume 90, pp. 3740 - 3743.

10. G. Gaussorgues (1989). *La thermographie infrarouge: principes, technologies, applications.* Technique et Documentation - Lavoisier, Paris, pp. 279 - 329.

11. J. Gordon (1990). Observations on efficiency of heat engines operating at maximum power. *American Journal of Physics*, volume 58, pp. 370 - 375.

12. A. De Vos (1990). De endoreversibele termodynamika: een nuttige aanvulling bij de formule van Carnot? *Het Ingenieursblad*, volume 59, number 9, pp. 62 - 70.

13. G. De Mey and G. Merckaert (1990). Infrarood termografie: ruime mogelijkheden voor industrie en geneeskunde. *Het Ingenieursblad*, volume 59, number 11, pp. 63 - 71.

4
Wind energy creation

4.1 Introduction

In this chapter, in contrast to Chapter 2, we will consider the possibility of temperature differences on the surface of the planets. For simplicity we will consider only two temperatures: part of the planet's surface is at the temperature T_3 and the rest is at the (lower) temperature T_4.

The existence of two reservoirs at different temperatures makes the creation of work possible. Indeed, the planet's atmosphere is a working fluid that converts heat into mechanical work. Above locations that are highly illuminated by the Sun, gases tend to rise in temperature and therefore to expand. The idler (and thus lighter) gas rises to the upper atmosphere. When it moves to less-illuminated locations, it cools, becomes heavier and tends to sink back to the planet's surface. Such macroscopic cycles are the origin of the winds.

Description of wind creation in terms of endoreversible thermodynamics was first introduced by Gordon and Zarmi [5], and was later refined by De Vos and Flater [7].

4.2 Preliminary model

Figure 4.1(a) shows the first model under consideration. We will call it the Gordon and Zarmi model. It consists of a planet, the sunny side of which is a heat reservoir at temperature T_3 and the dark side of which is a reservoir at temperature T_4. If we immediately neglect starlight, moonlight, and 2.74 K radiation with respect to the solar radiation, then the following energies are exchanged.

- The illuminated hemisphere
    - ∗ emits a power $2\pi R^2\sigma(1-\gamma)T_3^4$, and
    - ∗ receives a power $\pi R^2\sigma(1-\rho)fT_s^4$ from the Sun.

- The dark hemisphere
  - * emits a power $2\pi R^2\sigma(1-\gamma)T_4^4$, and
  - * receives no power from the night sky.

Thus we have the following two energy exchanges:

$$Q_1 = 2\pi R^2\sigma[\,(1-\rho)\frac{f}{2}T_s^4 - (1-\gamma)T_3^4\,] \tag{4.1}$$

$$Q_2 = 2\pi R^2\sigma\,(1-\gamma)T_4^4\,. \tag{4.2}$$

Maximum work can be extracted from the temperature difference between T_3 and T_4, if we interconnect the two reservoirs by means of a reversible heat engine. Reversibility is expressed by

$$\frac{Q_1}{T_3} = \frac{Q_2}{T_4} \tag{4.3}$$

or

$$\frac{(1-\rho)\frac{f}{2}T_s^4 - (1-\gamma)T_3^4}{T_3} = (1-\gamma)T_4^3\,. \tag{4.4}$$

If we denote by W the created work, then conservation of energy is expressed by

$$Q_1 = Q_2 + W \tag{4.5}$$

or

$$2\pi R^2\sigma\,[\,(1-\rho)\frac{f}{2}T_s^4 - (1-\gamma)T_3^4\,] = 2\pi R^2\sigma\,(1-\gamma)T_4^4 + W\,. \tag{4.6}$$

If we denote by η the efficiency of the engine, we have $W = \eta Q_1$. Combined with the energy conservation of eqn (4.5) and the reversibility condition of eqn (4.3), this yields the familiar

$$\eta = 1 - \frac{T_4}{T_3}\,. \tag{4.7}$$

Solving the set of two eqns (4.4) and (4.7) for T_3 and T_4 yields

$$T_3 = \frac{1}{[1+(1-\eta)^3]^{1/4}}\,2^{1/4}\,T_p$$

$$T_4 = \frac{1-\eta}{[1+(1-\eta)^3]^{1/4}}\,2^{1/4}\,T_p\,, \tag{4.8}$$

where T_p is the planet's surface temperature (2.6), according to the uniform-temperature model from Chapter 2:

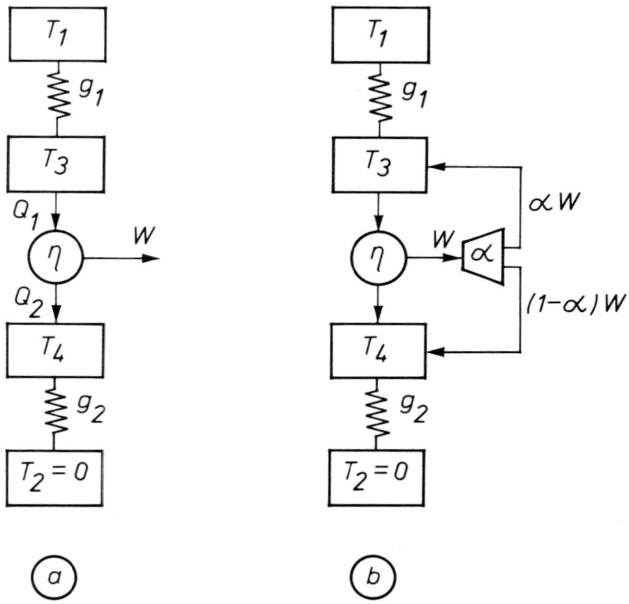

Fig. 4.1. Two models for the conversion of solar energy into wind energy: (a) preliminary model, (b) extended model.

$$T_p = \frac{(1-\rho)^{1/4}}{(1-\gamma)^{1/4}} \frac{f^{1/4}}{\sqrt{2}} T_s \ .$$

Substitution of the T_3 and T_4 results into eqn (4.6) gives

$$W = \frac{\eta(1-\eta)^3}{1+(1-\eta)^3} Q_s \ , \tag{4.9}$$

where Q_s denotes the total amount of solar energy absorbed by the planet:

$$\begin{aligned} Q_s &= \pi R^2 Z \\ &= \pi R^2 (1-\rho) S \\ &= \pi R^2 (1-\rho) f \sigma T_s^4 \ . \end{aligned}$$

We now define the dimensionless work

$$w = \frac{W}{Q_s} \ .$$

Because the quantity is a ratio of a work output to a heat input, it deserves the name efficiency. However, we have to be careful not to confuse with the efficiency

$$\begin{aligned}\eta &= \frac{W}{Q_1}\\ &= \frac{W}{Q_s - 2\pi R^2\sigma(1-\gamma)T_3^4}\ .\end{aligned}$$

Therefore, we will call w the 'solar energy efficiency', whereas we will call η the 'engine efficiency'. Both quantities are related. Indeed, the work-efficiency characteristic of eqn (4.9) can be written

$$w = \frac{\eta(1-\eta)^3}{1+(1-\eta)^3}\ . \tag{4.10}$$

Figure 4.2 shows the curve $w(\eta)$.

We now look for the η value, which maximizes the power W. The maximum is found by straightforward execution of $dw/d\eta = 0$. This yields

$$(1-\eta)^4 + 4(1-\eta) - 3 = 0\ . \tag{4.11}$$

This equation is nothing else but the eighth-degree Stefan–Boltzmann equation (3.20) for $T_2 = 0$. Indeed, the transport equations (4.1)-(4.2) can be rewritten as follows:

$$\begin{aligned}Q_1 &= 2\pi R^2\sigma(1-\gamma)\ (\ \frac{1-\rho}{1-\gamma}\ \frac{f}{2}\ T_s^4 - T_3^4\ )\\ Q_2 &= 2\pi R^2\sigma(1-\gamma)\ (\ T_4^4 - 0\ )\ ,\end{aligned}$$

which are the Stefan–Boltzmann transport equations

$$\begin{aligned}Q_1 &= g_1(T_1^4 - T_3^4)\\ Q_2 &= g_2(T_4^4 - T_2^4)\ ,\end{aligned}$$

with $g_1 = g_2$ and with

$$\begin{aligned}T_1 &= \frac{(1-\rho)^{1/4}}{(1-\gamma)^{1/4}}\ \frac{f^{1/4}}{2^{1/4}}\ T_s = 2^{1/4}\ T_p\\ T_2 &= 0\ .\end{aligned}$$

The result $T_2 = 0$ is of course a consequence of neglecting the incidence of starlight, moonlight, and cosmic radiation on the dark hemisphere. In contrast to the general equation (3.20), eqn (4.11) has an analytical solution,

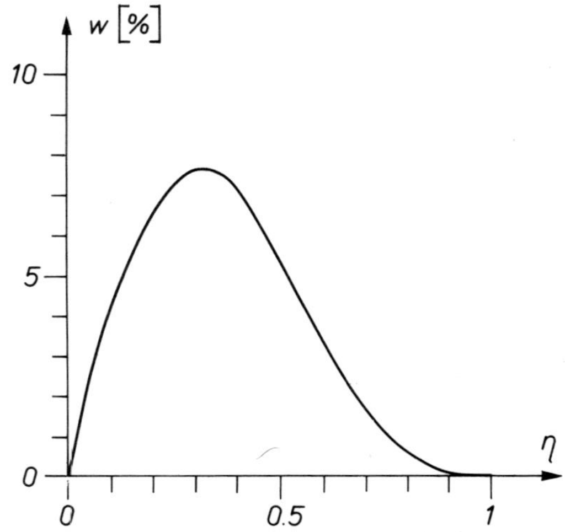

Fig. 4.2. Produced power w as a function of engine efficiency η.

i.e. a solution in radicals. See Appendix D. However, the expression for η is so horribly complicated, we prefer to use the numerical result. The one and only solution of eqn (4.11) satisfying $0 < \eta < 1$ is $\eta = 0.30749$. The approximate value 0.31 for this maximum-power efficiency η can be read from Fig. 3.9 of Chapter 3 (curve labelled $g_1 = g_2$, at abscissa $T_2/T_1 = 0$). Substitution of $\eta = 0.30749$ into eqns (4.8) and (4.9) yields

$$\begin{aligned} T_3 &= 1.317\ T_p \\ T_4 &= 0.912\ T_p \\ W &= 0.0767\ Q_s\ . \end{aligned}$$

Thus 7.67 per cent of the solar energy Q_s can be converted into work, e.g. into wind energy. Note that this solar energy conversion efficiency $w = 0.0767$ is independent of f, ρ and γ and thus is the same for all planets. It is even independent of T_s and is thus the same for all solar systems.

4.3 Final model

The above simple model has one important drawback: it is correct only if the mechanical power W is either continuously stored or continuously drained away from the planet. In reality neither of these two processes happen: the wind energy W is dissipated as heat in the same planet (after a temporary storage in the atmosphere). Therefore, we have to complete the Gordon and Zarmi model of Fig. 4.1(a) with the deposition of an amount of heat, equal to W. We assume, after De Vos and Flater [7], that a fraction α is deposited on the illuminated side of the planet and the rest is deposited on the dark side. See Fig. 4.1(b). The dissipation distribution factor α of course satisfies

$$0 \leq \alpha \leq 1 .$$

Now the reversibility expression (4.4) of the conversion engine is replaced by

$$\frac{2\pi R^2\sigma[\,(1-\rho)\,\frac{f}{2}\,T_s^4-(1-\gamma)T_3^4\,]+\alpha W}{T_3} = \frac{2\pi R^2\sigma(1-\gamma)T_4^4-(1-\alpha)W}{T_4} \,. \tag{4.12}$$

Eqn (4.6) expressing the conservation of energy is replaced by

$$2\pi R^2\sigma\;[\,(1-\rho)\,\frac{f}{2}\;T_s^4-(1-\gamma)T_3^4\,] = 2\pi R^2\sigma\;(1-\gamma)T_4^4 \;. \tag{4.13}$$

We still have

$$\eta = 1-\frac{T_4}{T_3} \;. \tag{4.14}$$

Solving the set of equations (4.13)-(4.14) for T_3 and T_4 yields

$$\begin{aligned} T_3 &= \frac{1}{[1+(1-\eta)^4]^{1/4}}\;2^{1/4}\;T_p \\ T_4 &= \frac{1-\eta}{[1+(1-\eta)^4]^{1/4}}\;2^{1/4}\;T_p \;. \end{aligned} \tag{4.15}$$

Substitution of these results into eqn (4.12) leads to W, and thus, after division by Q_s, to w:

$$w = \frac{\eta(1-\eta)^4}{(1-\alpha\eta)[1+(1-\eta)^4]} \;.$$

This equation replaces eqn (4.10). It contains, besides the parameter η, the parameter α. Figure 4.3 shows curves $w(\eta)$ for different values of α. We

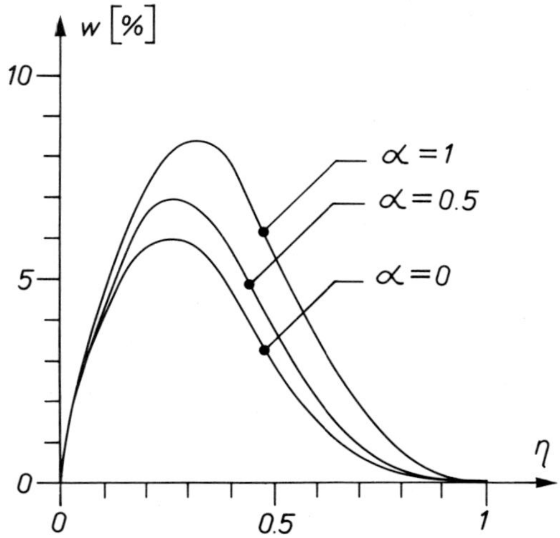

Fig. 4.3. Produced power w as a function of engine efficiency η and dissipation distribution factor α.

see that maximum wind energy is created if $\alpha = 1$, i.e. if all wind energy is dissipated on the illuminated side of the planet. In that case, we have

$$w = \frac{\eta(1-\eta)^3}{1+(1-\eta)^4} \ . \tag{4.16}$$

For this particular value of α, maximization with respect to η, i.e. $dw/d\eta = 0$, yields again

$$(1-\eta)^4 + 4(1-\eta) - 3 = 0 \tag{4.17}$$

or $\eta = 0.30749$. Substitution into eqn (4.16) gives

$$w = 0.0830 \ .$$

Thus 8.30 per cent of the solar energy can be converted into wind energy. With the assumption $\alpha = 0.5$, we get 6.90 per cent. Finally, for $\alpha = 0$, we get 6.01 per cent. Again these percentages are not affected by the parameters T_s, f, ρ, and γ.

It is interesting to note from eqn (4.15) that the two surface temperatures T_3 and T_4 satisfy

$$\frac{T_3^4 + T_4^4}{2} = T_p^4 \ ,$$

such that we can say that the unique temperature T_p, found in the uniform-planet model of Section 2.2, is some kind of average of the two temperatures

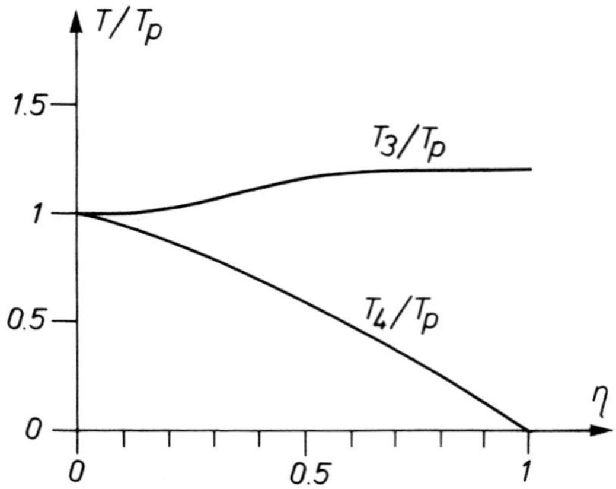

Fig. 4.4. Temperatures T_3 and T_4 as a function of the engine efficiency η.

T_3 and T_4, found in this section. Figure 4.4 shows how T_3 and T_4 vary with the parameter η.

4.4 Why 8 per cent ? ♡

One can ask why the fraction of the solar energy input, converted into work, is maximum 8.3 per cent? The number turns out to be correlated with the power of T in the Stefan–Boltzmann law and thus with the dimensionality of our universe.

For an n-dimensional planet, conservation of entropy in the reversible part of the engine is expressed as

$$\frac{Q_s - (1-\gamma)A\ T_3^{n+1} + \alpha W}{T_3} = \frac{(1-\gamma)A\ T_4^{n+1} - (1-\alpha)W}{T_4}$$

and conservation of energy is expressed by

$$Q_s - (1-\gamma)A\ T_3^{n+1} = (1-\gamma)A\ T_4^{n+1} \ .$$

Here Q_s denotes again the total solar energy absorbed by the planet, whereas A denotes the n-dimensional Stefan–Boltzmann constant σ_n times half the surface area of the planet. Finally, we still have

$$\eta = 1 - \frac{T_4}{T_3} \ .$$

Table 4.1. Solution of the equation $(1-\eta)^{n+1}+(n+1)(1-\eta)-n=0$.

n	η	w
1	0.586	0.2071
2	0.404	0.1184
3	0.307	0.0830
4	0.248	0.0639
5	0.208	0.0520
6	0.179	0.0438
7	0.157	0.0378
8	0.140	0.0333
9	0.126	0.0298
10	0.115	0.0269
11	0.105	0.0245
12	0.097	0.0225
...		
25	0.049	0.0110
...		

Elimination of T_3 and T_4 from the above three equations yields

$$w=\frac{\eta(1-\eta)^{n+1}}{(1-\alpha\eta)[1+(1-\eta)^{n+1}]} .$$

For $\alpha=1$ this becomes

$$w=\frac{\eta(1-\eta)^{n}}{1+(1-\eta)^{n+1}} , \tag{4.18}$$

the generalization of eqn (4.16). Applying $dw/d\eta=0$ now yields an equation of $(n+1)^{th}$ degree:

$$(1-\eta)^{n+1}+(n+1)(1-\eta)-n=0 ,$$

the generalization of eqn (4.17). Table 4.1 gives the numerical solution of this equation for various values of n.

Substitution of the solution η into eqn (4.18) yields the value of w, also tabulated in Table 4.1. We see that the solar energy efficiency w depends strongly on the number of dimensions. For large n, it diminishes as $1/n$.

4.5 Tidal winds and zonal winds

The winds modelled in the previous sections are called tidal winds or thermal tides, because the air displacements resemble closely the water displacements of the sea tides. We recall here that they are caused by longitudinal temperature differences: the diurnal longitudes are hotter than the nightly longitudes.

The most important wind structures, however, are the zonal winds or Hadley circulations [3] [4]. Those are caused by latitudinal temperature differences: the equatorial latitudes are hotter than the polar latitudes. Indeed, because the Sun is closer to the zenith, more solar energy falls on a unit surface area of the equatorial zone than on a unit surface area of the (ant)arctic.

In contrast to the tidal wind model, described in Sections 4.2 - 4.4, the zonal wind model completely neglects longitudinal temperature gradients. This means that the planet's rotation suffices to eliminate any temperature difference between the illuminated hemisphere of the planet and the dark hemisphere of the planet. See Fig. 4.5.

We will assume, for sake of simplicity, that the inclination of the equator to the ecliptic (i.e. the plane of the planet's orbit around the Sun) is zero. With this symmetry, we can divide the planet into three parts: one equatorial (or tropic) zone, expanding from latitude $-\varphi$ to latitude φ, and two polar (or arctic) zones, one north and one south of the tropic zone. See Fig. 4.6. The equatorial zone is assumed to have a uniform surface temperature T_3, whereas the two polar zones are assumed to have a uniform temperature T_4. Again neglecting any background radiation (stars, moons, cosmic), we have the following energy exchanges:

- The equatorial zone is illuminated by

$$A_1 Z = A_1(1-\rho) f \sigma T_s^4 \ .$$

- The equatorial zone emits

$$A_3(1-\gamma)\sigma T_3^4 \ .$$

- The polar zones are illuminated by

$$A_2 Z = A_2(1-\rho) f \sigma T_s^4 \ .$$

- The polar zones emit

$$A_4(1-\gamma)\sigma T_4^4 \ .$$

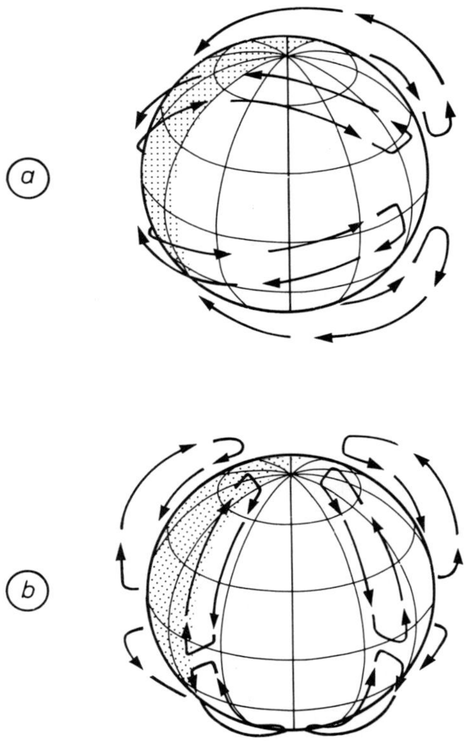

Fig. 4.5. Two wind models: (a) tidal winds, (b) zonal winds.

Here we introduce the short-hand notations A_1, A_2, A_3 and A_4:

- A_3 is the surface area of the equatorial zone,
- A_1 is the area of the projection of this surface on a plane perpendicular to the Sun's rays, i.e. perpendicular to the line Sun–planet,
- A_4 is the total surface area of the two polar zones, and
- A_2 is the area of the projection of this surface on a plane perpendicular to the Sun's rays.

Because of our simplifying assumption that the Sun lies in the equator, we have the following simple trigonometric expressions:

$$
\begin{aligned}
A_1 &= R^2[\ 2\varphi + \sin(2\varphi)\ ] \\
A_3 &= 4\pi R^2 \sin(\varphi) \\
A_2 &= R^2[\ \pi - 2\varphi - \sin(2\varphi)\ ] \\
A_4 &= 4\pi R^2[\ 1 - \sin(\varphi)\ ]\ .
\end{aligned}
$$

The above energy exchanges are summarized by the following Stefan–Boltzmann transport equations:

$$
\begin{aligned}
Q_1 &= A_3\ (1-\gamma)\sigma\ (\ \frac{A_1}{A_3}\,\frac{1-\rho}{1-\gamma}\ fT_s^4 - T_3^4\ ) \\
Q_2 &= A_4\ (1-\gamma)\sigma\ (\ T_4^4 - \frac{A_2}{A_4}\,\frac{1-\rho}{1-\gamma}\ fT_s^4\ )\ .
\end{aligned}
$$

We immediately recognize the general form

$$
\begin{aligned}
Q_1 &= g_1(T_1^4 - T_3^4) \\
Q_2 &= g_2(T_4^4 - T_2^4)\ .
\end{aligned}
$$

In contrast to the tidal wind model, we have now $g_1 \neq g_2$ and $T_2 \neq 0$. Therefore we can predict that the zonal wind model leads to more complicated calculations. There is even a supplementary cause of complication: which value do we have to choose for φ, the demarcation line between tropic and arctic zones? The most elegant choice is the one that maximizes the work production W. Thus maximum-power production is then defined by the set of conditions $\partial W/\partial\eta = 0$ and $\partial W/\partial\varphi = 0$.

We restrict ourselves here to the results of the calculations: Fig. 4.7 shows (for the dissipation distribution factor $\alpha = 1$) the dimensionless wind power generation w as a function of both φ and η. The *maximum maximorum* amounts to $w = 0.0042$ and occurs at $\eta = 0.082$, for a boundary at $\varphi = 49.7\ ^\circ$.

Thus the zonal model, where all day–night temperature differences are neglected, gives rise to a maximum of 0.42 per cent solar energy conversion, a far less efficient conversion than the tidal model, where all latitude temperature differences are neglected. However, we have to face the fact that the Sun is in the equator only two days a year (i.e. on the two equinoxes). If we introduce a non-zero declination of the Sun, calculations become much harder, especially because the expressions for the capture cross-sections A_1 and A_2 become complicated. Numerical calculations result in higher values of w, i.e. in the order of magnitude of a few per cent [8].

4.6 Two examples

We summarize that the Gordon–Zarmi model leads to fundamental upper limits for the conversion of solar energy into wind, valid for any planet from

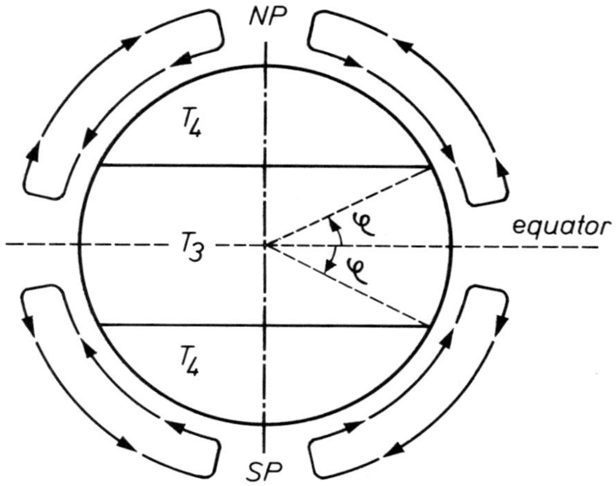

Fig. 4.6. Zonal wind model.

any solar system. The upper limit of the wind heat engine is 8 per cent, a universal value independent of key parameters.

Because of thermal inertia, combined with the planet's rotation around its axis, day–night temperature differences are smoothed out, such that actual conversion efficiencies are smaller than the upper limit of 8 per cent.

We will discuss two examples: Mars and the Earth. The reason is simple: too little is known (yet) about the conversion efficiency of solar energy into wind energy in the atmosphere of any other planet.

Whereas tidal winds play a substantial role in the atmosphere of Mars, they are of lesser importance for the Earth. Zonal wind patterns occur in both planets. However, in each case the number of zones differs from three and thus the number of Hadley cells differs from two:

- In the atmosphere of Mars only two zones (and thus only one Hadley cell) are present, because of the planet's inclination combined with the absence of oceans. This causes a summer hemisphere at high temperature and a winter hemisphere at low temperature.
- The atmosphere of the Earth, on the other hand, shows seven temperature zones (and thus six Hadley cells or wind belts), because of the combined effect of the thermal capacity of the Earth's oceans and the Coriolis forces caused by the Earth's rotation.

See Fig. 4.8.

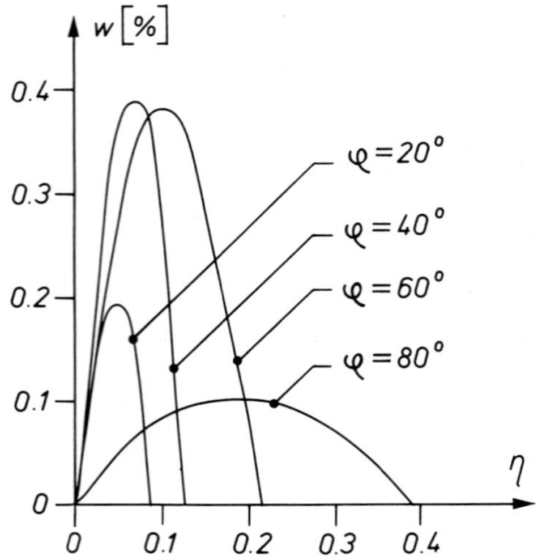

Fig. 4.7. Produced power w as a function of engine efficiency η and interzone boundary latitude φ.

Experimental data [2] for the Earth give about 3 per cent for w. Numerical simulations [1] for Mars lead to an estimation for w an order of magnitude smaller than for the Earth: about 0.2 per cent.

Bibliography

1. C. Leovy and Y. Mintz (1969). Numerical simulation of the atmospheric circulation and climate of Mars. *Journal of Atmospheric Sciences*, volume 26, pp. 1167 - 1190.

2. M. Gustavson (1979). Limits to wind power utilization. *Science*, volume 204, pp. 13 - 17.

3. J. Beatty, B. O'Leary and A. Chaikin (1981). *The new solar system.* Cambridge University Press, Cambridge, pp. 57 - 70 and 117 - 128.

4. F. Shu (1982). *The physical universe, an introduction to astronomy.* University Science Books, Mill Valley, pp. 433 - 458.

5. J. Gordon and Y. Zarmi (1989). Wind energy as a solar-driven heat engine: a thermodynamic approach. *American Journal of Physics*, volume 57, pp. 995 - 998.

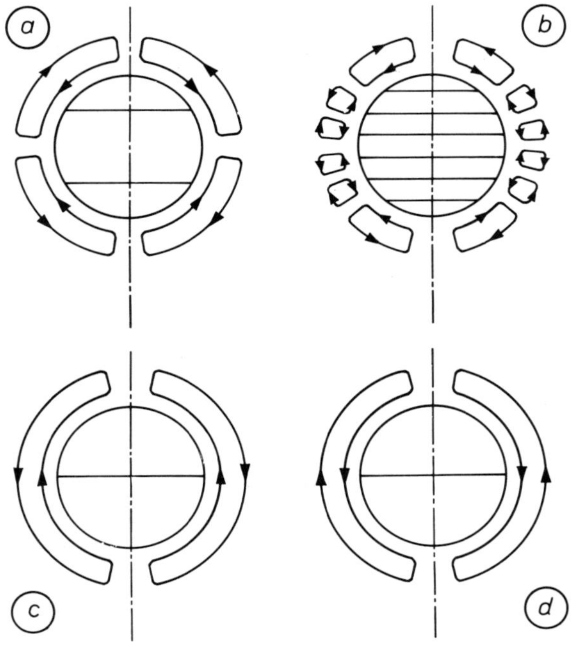

Fig. 4.8. Zonal winds: (a) simple Hadley model, (b) Earth model, (c) Mars model in summertime, (d) Mars model in wintertime.

6. L. Sertorio (1991). *Thermodynamics of complex systems.* World Scientific, Singapore.

7. A. De Vos and G. Flater (1991). The maximum efficiency of the conversion of solar energy into wind energy. *American Journal of Physics*, volume 59, pp. 751 - 754.

8. A. De Vos and P. van der Wel (1992). The efficiency of the conversion of solar energy into wind energy by means of Hadley cells. In press.

5
Photothermal conversion

5.1 Introduction

The solar energy conversion in the previous chapter had two special features:

- it is natural: no human intervention occurs; and
- it is global: it concerns the planet as a whole.

In this chapter we will consider a solar energy conversion system built by humans. As the human race does not (yet) influence the planet as a whole, the conversion will be local, i.e. concerning only a small part of the planet. Therefore the temperature of the planet's surface is not affected. In Chapters 2 and 4, the planet's surface temperature was an unknown of the problem. In this chapter, as well as in the following ones, the planet's temperature will be considered as a fixed and known number. We will choose the value T_p of Chapter 2, which is also 'a kind of average' of the two values T_3 and T_4 calculated in Chapter 4. The local nature of the converter also has another consequence: the surface area of the planet is no longer of any concern and all calculations will be done for a unit of surface area.

The local nature of the conversion process makes the model simple, as the dissipation of the produced work W happens on a different place than its generation. Therefore, we don't have to include any dissipation into the model and we don't have to bother about any dissipation distribution factor α as in Chapter 4.

We assume that technological skill is so good that the conversion device makes a perfect thermal contact with the planetary surroundings: $g_2 = +\infty$. Therefore $T_4 = T_2 = T_p$. The resulting configuration is shown in Fig. 5.1 and corresponds to the model introduced by Müser [1] in 1957. The solar converter is assumed to be oriented optimally, i.e. perpendicular to the Sun's rays. Therefore the incident radiation is equal to the solar constant $S = f\sigma T_s^4$, e.g. 1353 W/m^2 for the Earth.

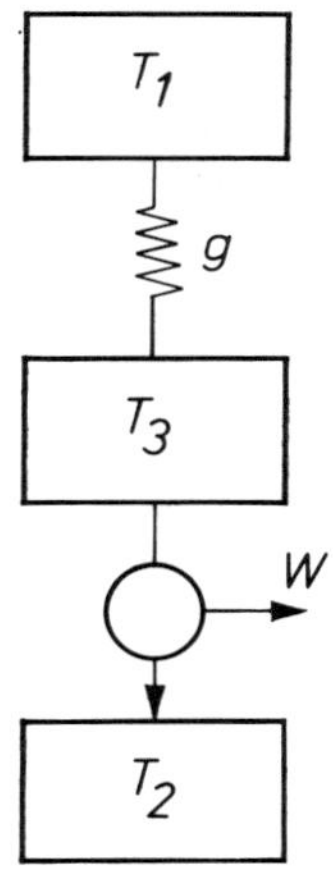

Fig. 5.1. Müser engine.

From all directions, not covered by the solar disk, the converter receives black-body radiation at temperature T_p: from the ground, from the surrounding buildings and/or trees and from the clouds and/or atmosphere. The converter itself radiates a black-body radiation of temperature T_3, called the converter temperature. See Fig. 5.2. Per unit of surface area it emits a power σT_3^4. Thus we assume (for simplicity) that only the upper surface emits radiation. The bottom side of the device has a zero emissivity ϵ, which can be realised, for example, by applying a mirror-like metal coating. If the bottom were not metallized, the body would emit $2\sigma T_3^4$, as a 1 m^2 flat radiator has 2 m^2 surface area.

The net total incident energy is given by

$$\begin{aligned} Q_1 &= f\sigma T_s^4 + (1-f)\sigma T_p^4 - \sigma T_3^4 \\ &= \sigma[\, f\,T_s^4 + (1-f)\,T_p^4 - T_3^4\,]\,. \end{aligned} \tag{5.1}$$

Equation (5.1) and Fig. 5.2 can be interpreted as follows. The converter radiates the full Stefan–Boltzmann flux σT_3^4 towards the hemisphere above it. It receives T_s^4 radiation only from a small solid angle ω. The incident solar radiation is therefore only a fraction f of the Stefan–Boltzmann flux σT_s^4. The aperture factor f is given by

$$f = \frac{\iint_\omega \cos\vartheta\, d\omega}{\iint_{2\pi} \cos\vartheta\, d\omega}\,.$$

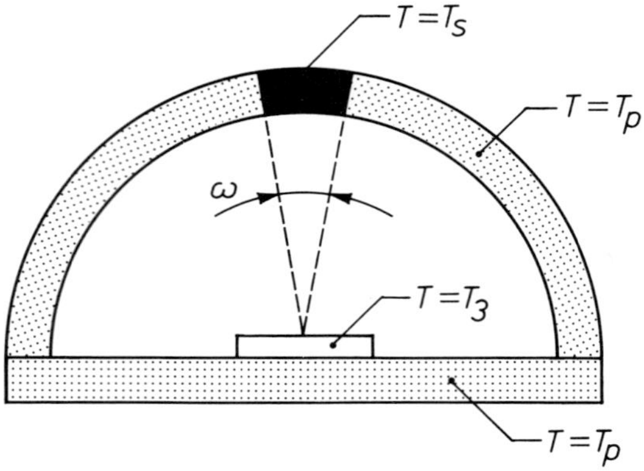

Fig. 5.2. Müser engine.

The denominator of this expression equals π; the numerator equals approximately ω, if the solar disc is small, as then the cosine of ϑ is almost unity in the whole solid angle ω. Thus

$$f \approx \frac{\omega}{\pi} .$$

Because ω equals πR_s^2 divided by r^2, we finally get

$$f \approx \frac{R_s^2}{r^2} ,$$

i.e. the dilution factor (2.2) from Chapter 2. Note that f was introduced in Chapter 2 from a 'solar' point of view: only a fraction of the Sun's rays reaches the planet, whereas here f is deduced from a 'planetary' point of view: the planet is illuminated by the Sun's rays from only a limited solid angle. Both approaches give the same result: the same correction factor f, much smaller than unity. What the rest of the hemisphere is concerned, it sends to the converter a Stefan–Boltzmann radiation σT_p^4 with aperture factor

$$\frac{\iint_{2\pi-\omega} \cos\vartheta \, d\omega}{\iint_{2\pi} \cos\vartheta \, d\omega} ,$$

i.e. $1 - f$.

We now compare the energy flow (5.1) with the general transport equation of the Stefan–Boltzmann machine, i.e. with $g_1(T_1^4 - T_3^4)$. We have

$$\begin{aligned} g_1 &= \sigma \\ T_1 &= [\, f\, T_s^4 + (1-f)\, T_p^4 \,]^{1/4} . \end{aligned} \tag{5.2}$$

The latter equation can be interpreted as follows. The solar absorber sees a hemisphere with non-uniform temperature. For the Stefan–Boltzmann engine however, this situation is equivalent to the case where the solar absorber is irradiated by a 'sky' at uniform temperature T_1. This temperature is 'some' weighted average of T_s and T_p.

As anticipated in Section 2.7, we will neglect here, for simplicity, the planet's albedo and greenhouse effects. Therefore, we assume that eqn (2.6) for the planet temperature T_p can be approximated by the simpler formula (2.5):

$$T_p = \frac{f^{1/4}}{\sqrt{2}}\, T_s \;.$$

We have seen in Table 2.1 that the approximation is good, except for Venus. Substitution into eqn (5.2) gives

$$T_1 = (\frac{5f}{4} - \frac{f^2}{4})^{1/4}\, T_s \;.$$

Because for any planet $f \ll 1$, we can approximate T_1 as follows:

$$T_1 \approx (\frac{5f}{4})^{1/4}\, T_s$$

or

$$T_1 \approx 5^{1/4}\, T_p \;.$$

Thus, the effective sky temperature T_1 equals 1.49535 times the planet temperature. For the Earth we have $T_p = 288$ K and thus $T_1 = 431$ K.

5.2 Solar energy efficiency

The photothermal converter produces an amount W of power per unit of time and per unit of surface area. Just as in Chapter 4, we introduce a solar energy conversion efficiency:

$$w = \frac{W}{f\sigma T_s^4} \;.$$

Again, this quantity should not be confused with the engine efficiency η, which is here given by

$$\eta = \frac{W}{f\sigma T_s^4 + (1-f)\sigma T_p^4 - \sigma T_3^4} \;,$$

where the net input heat Q_1 figures in the denominator, whereas in the definition of w the incident solar heat figures in the denominator. Thus, in the denominator of w:

- the flux $(1-f)\sigma T_p^4$ from the surroundings is not considered as input, and
- the flux σT_3^4 emitted by the absorbing plate is not considered as a correction to the input.

We will now calculate W and thus also its dimensionless form w. From

$$\begin{aligned} W &= \eta Q_1 \\ Q_1 &= f\sigma T_s^4 + (1-f)\sigma T_p^4 - \sigma T_3^4 \\ T_3 &= \frac{T_4}{1-\eta} \\ T_4 &= T_p \ , \end{aligned}$$

we can deduce immediately that

$$W = \eta \ [\ f\sigma T_s^4 + (1 - f - \frac{1}{(1-\eta)^4})\sigma T_p^4 \ ]$$

and thus

$$w = \eta \ [\ 1 + \frac{(1-f)(1-\eta)^4 - 1}{(1-\eta)^4} \ \frac{T_p^4}{fT_s^4} \ ] \ .$$

With the approximation

$$T_p = \frac{f^{1/4}}{\sqrt{2}} \ T_s \ ,$$

this eventually leads to the relationship between w and η, we were searching for:

$$w = \eta \ \frac{(5-f)(1-\eta)^4 - 1}{4(1-\eta)^4} \ . \tag{5.3}$$

Note that with very good accuracy this can be approximated by

$$w \approx \eta \ \frac{5(1-\eta)^4 - 1}{4(1-\eta)^4} \ . \tag{5.4}$$

Figure 5.3 shows this function $w(\eta)$.

Note that for

$$\eta > 1 - \frac{1}{5^{1/4}} \ ,$$

we have negative $w(\eta)$, because $T_3 > T_1$ and therefore the heat current $Q_1 = g_1(T_1^4 - T_3^4)$ changes direction. The device then works as a heat pump, instead of as a heat engine.

Between its two zeros, i.e. between $\eta = 0$ and $\eta = \eta_{oc} = 1 - 5^{-1/4} = 0.33126$, the function $w(\eta)$ displays a maximum. The coordinates of this extremum are determined by solving $dw/d\eta = 0$, i.e. by solving

$$5(1-\eta)^5 + 3(1-\eta) - 4 = 0 \ . \tag{5.5}$$

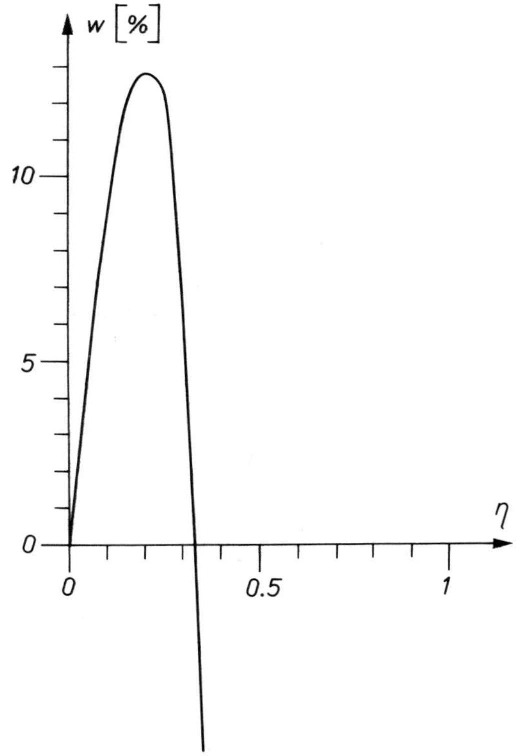

Fig. 5.3. Solar energy conversion efficiency w as a function of engine efficiency η of Müser engine.

5.3 The Müser engine

Before commenting on the result of eqn (5.5), let us relate the calculations of Section 5.2 to the general theory of the Stefan–Boltzmann engines. The Müser engine is a man-made Stefan–Boltzmann engine as described in Section 5.1. This means that the reservoir 4 is in perfect thermal contact with reservoir 2:

$$T_4 = T_2 \ .$$

Only the heat flow Q_1 to the Carnot engine is limited by a finite conductance g_1 :

$$Q_1 = g_1(T_1^4 - T_3^4) \ . \tag{5.6}$$

From the previous discussion on Stefan–Boltzmann engines we know that maximum power is generated for

$$T_1^4(1-\eta)^5 + 3T_2^4(1-\eta) - 4T_2^4 = 0 \ . \tag{5.7}$$

As η equals $1 - T_4/T_3$ and thus equals $1 - T_2/T_3$, this equation can also be written as

$$4T_3^5 - 3T_2T_3^4 - T_1^4T_2 = 0 \ .$$

This equation was discovered by Müser [1] in 1957, and afterwards rediscovered independently by Castañs [5] in 1976, by Jeter [8] and by De Vos and Pauwels [9] in 1981.

As stated in Chapter 3, eqn (5.7) has no analytical solution. The numerical solution can be seen as the curve labelled $g_1/g_2 = 0$ of Fig. 3.9.

If we now apply $T_1 = 5^{1/4}T_p$ and $T_2 = T_p$, then eqn (5.7) becomes the quintic equation (5.5).

Whereas the solutions of a fourth-degree equation can always be written in terms of radicals, the solutions of an arbitrary polynomial equation of degree 5 cannot, in general. Only some special classes of fifth-degree equations give rise to analytical solutions in terms of nested radicals. Equation (5.5), although of simple appearance, is not a member of such a special class. Mathematicians say that it belongs to the wrong Galois group. Therefore we have to appeal to numerical methods for solving it. This yields

$$\eta = 0.20292 \ .$$

Subsequent substitution into eqn (5.4) gives

$$w = 0.12797 \ .$$

Thus a Müser engine has an efficiency that is independent of f and therefore has a universal value, i.e. 12.8 per cent. This efficiency is higher than the efficiency of the wind engine of Chapter 4. The reason is mainly the fact that the wind engine suffers from two conduction limitations g_1 and g_2, whereas the Müser engine only suffers from one finite conductance g_1.

5.4 Why 13 per cent ? ♡

As in Section 4.4, one can ask why the fraction of the solar energy input converted into work is limited to 13 per cent? The number again turns out to be correlated with the power of T in the Stefan–Boltzmann law and thus with the dimensionality of our universe.

Before we can construct an n-dimensional equivalent of eqn (5.5), we have to construct the n-dimensional generalization of eqn (2.5), i.e. of

$$T_p = (\frac{1}{4} f)^{\frac{1}{4}} T_s .$$

The reader will immediately guess that the generalization of the exponent $\frac{1}{4}$ is given by $\frac{1}{n+1}$. The generalization of the factor $\frac{1}{4}$ in front of f is, however, less staightforward. Indeed, the thermal balance of the planet

$$\sigma \pi R^2 f T_s^4 = \sigma 4\pi R^2 \dot{T}_p^4$$

becomes, in n dimensions,

$$\sigma_n V_{n-1} R^{n-1} f T_s^{n+1} = \sigma_n S_n R^{n-1} T_p^{n+1} .$$

Here $S_n R^{n-1}$ is the surface area of the planet, whereas $V_{n-1} R^{n-1}$ is the surface area of its projection on a hyperplane perpendicular on the line Sun–planet. Indeed, the projection of an n-dimensional hypersphere is an $(n-1)$-dimensional hypersphere, its $(n-1)$-dimensional volume being a surface area from an n-dimensional point of view.

The above thermal balance gives rise to

$$F_n T_p^{n+1} = f T_s^{n+1} ,$$

where F_n denotes the ratio of the surface area of a hypersphere to the surface area of its projection:

$$F_n = \frac{S_n}{V_{n-1}} .$$

In Appendix B the values of F_n are tabulated. Thus finally we get the following expression for the planet temperature:

$$T_p = (\frac{1}{F_n} f)^{\frac{1}{n+1}} T_s .$$

For $n = 3$ we have $F_3 = 4$, so that the old formula for T_p is recovered.

In a second step, we have to calculate the equivalent temperature T_1 of the n-dimensional sky. This is immediately found from

$$f T_s^{n+1} + (1-f) T_p^{n+1} = T_1^{n+1}$$

and thus is given by

$$T_1 = [\ fT_s^{n+1} + (1-f)T_p^{n+1}\ ]^{\frac{1}{n+1}} \ .$$

or

$$T_1 \approx (\frac{F_n + 1}{F_n}\ f)^{\frac{1}{n+1}}\ T_s \ .$$

Now, from the generalized, i.e. n-dimensional, Stefan–Boltzmann transport equation

$$Q_1 = \sigma_n (T_1^{n+1} - T_3^{n+1})$$

we get the $W(\eta)$ characteristic:

$$W = \eta\ \sigma_n\ [\ T_1^{n+1} - \frac{T_2^{n+1}}{(1-\eta)^{n+1}}\ ] \ .$$

With the above value of T_1 and with T_2 replaced by the above value of T_p, we get, after division by $\sigma_n f T_s^{n+1}$

$$w = \eta\ \frac{(F_n + 1)(1-\eta)^{n+1} - 1}{F_n(1-\eta)^{n+1}} \ . \tag{5.8}$$

Maximizing with respect to η yields the following maximum power condition:

$$(F_n + 1)(1-\eta)^{n+2} + n\ (1-\eta) - (n+1) = 0 \ , \tag{5.9}$$

the generalization of eqn (5.5) searched for. Table 5.1 gives the numerical solution of this equation for various values of n. For large n, it diminishes, somewhat more slowly than $1/n$.

Substitution of the solution η into eqn (5.8) yields the value of w, also tabulated in Table 5.1. We see how solar energy efficiency goes down with increasing dimensionality.

5.5 Concentrators

Taking into account human ingenuity, we assume that the incident radiation can be artificially augmented by a factor C (larger than unity), called the concentration factor. This is realized by use of optical devices, i.e. either by lenses (refractive devices) or by mirrors (reflective devices) or by a combination of both.

It is, however, important to note that human technology is limited by physical laws: one cannot make C infinitely large. One cannot focus the Sun's rays into an infinitely small area. The highest concentration is realized by capturing all rays leaving a unit of Sun surface area and making an optical image of it on a unit of surface area on the planet. This would require a huge elliptic mirror with the Sun at one focal point and the planet

Table 5.1. Solution of equation (5.9).

n	η	w
1	0.253	0.153
2	0.230	0.143
3	0.203	0.128
4	0.181	0.115
5	0.163	0.105
6	0.149	0.096
7	0.137	0.089
8	0.127	0.083
9	0.119	0.076
10	0.112	0.073
11	0.105	0.069
12	0.100	0.066
...		
25	0.060	0.041
...		

at the other focal point. See Fig. 5.4. The resulting power incident on the solar energy converter would equal the solar energy radiation density at the Sun's surface, i.e. σT_s^4. Thus

$$C_{max} f \sigma T_s^4 = \sigma T_s^4 \ .$$

Therefore the highest concentration C_{max} is $1/f$. When $C = C_{max}$, the solar absorber 'sees' the solar disc in a 2π solid angle (see Fig. 5.5) instead of in a limited solid angle (as in Fig. 5.2). Note that the highest concentration factor $1/f$ equals 46 300 for the Earth.

For an arbitrary C, the incident solar radiation is $Cf\sigma T_s^4$ with

$$1 \leq C \leq \frac{1}{f} \ . \tag{5.10}$$

From all directions, except those covered by the lenses and/or mirrors, the converter still receives radiation from the surroundings at temperature T_p. The black-body radiator still emits a power σT_3^4 per unit of surface area . The net total incident energy is now given by

$$\begin{aligned} Q_1 &= Cf\sigma T_s^4 + (1 - Cf)\sigma T_p^4 - \sigma T_3^4 \\ &= \sigma[\ Cf\, T_s^4 + (1 - Cf)\, T_p^4 - T_3^4\ ] \ , \end{aligned} \tag{5.11}$$

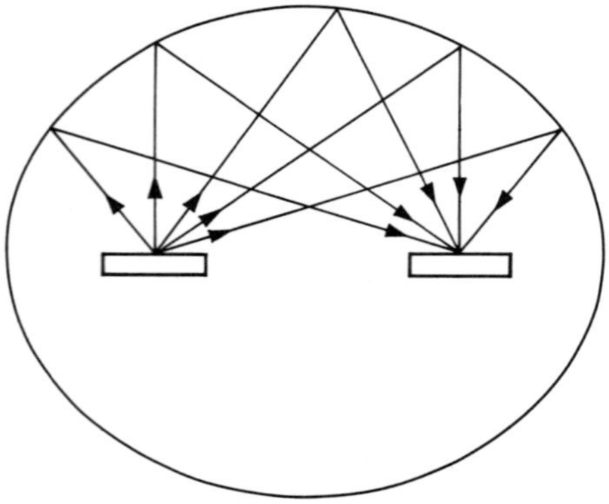

Fig. 5.4. Ideal solar energy concentrator.

such that

$$T_1 = [\ Cf\, T_s^4 + (1 - Cf)\, T_p^4\ ]^{1/4}\ .$$

This effective 'sky' temperature is higher than the value $5^{1/4}T_p$ of the Müser engine without concentrator. Indeed, with eqn (5.10) we can prove

$$5^{1/4}T_p \le T_1 \le T_s\ .$$

The upper limit T_s is reached for the extreme concentration factor $C = 1/f$. Thus, depending on the concentration factor C, the effective temperature

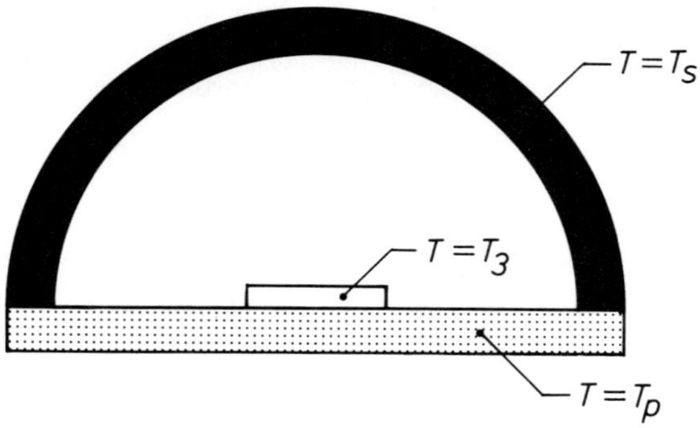

Fig. 5.5. Müser engine with ideal solar energy concentrator.

T_1 ranges from 1.49535 times the planet temperature to the Sun temperature.

We again have to define a solar energy efficiency, as in Section 5.2:

$$w = \frac{W}{Cf\sigma T_s^4} .$$

After some calculations, we find the generalization of eqn (5.3):

$$w = \eta \, \frac{(4C + 1 - Cf)(1-\eta)^4 - 1}{4C(1-\eta)^4} .$$

Note that for fully concentrated sunlight (i.e. $C = 1/f$)

$$w = \eta \, [\, 1 - \frac{f}{4(1-\eta)^4} \,] . \tag{5.12}$$

This function is zero for $\eta = 0$ and for $\eta = \eta_{oc} = 1 - f^{1/4}/\sqrt{2}$. In between, it has a maximum for η satisfying $dw/d\eta = 0$, i.e. satisfying

$$(1-\eta)^5 + \frac{3}{4} f(1-\eta) - f = 0 . \tag{5.13}$$

5.6 The Müser engine with concentrator

In contrast with the Müser engine with $C = 1$, the theory of the Müser engine with $C \gg 1$ can take advantage of the fact that $T_2 \ll T_1$. This leads to an approximate analytical solution for eqn (5.7). It can be found by neglecting the second term of the left-hand side of the equation, yielding

$$T_1^4 (1-\eta)^5 - 4T_2^4 \approx 0$$

or

$$\eta \approx 1 - 4^{1/5} \, (\frac{T_2}{T_1})^{4/5} .$$

For example, for the special case of fully concentrated sunlight, i.e. $Cf = 1$, we have $T_2 = T_p$ and $T_1 = T_s = 5762$ K, so that indeed $T_2 \ll T_1$. We have

$$\frac{T_2}{T_1} = \frac{T_p}{T_s} = \frac{f^{1/4}}{\sqrt{2}}$$

and thus

$$\eta \approx 1 - f^{1/5} .$$

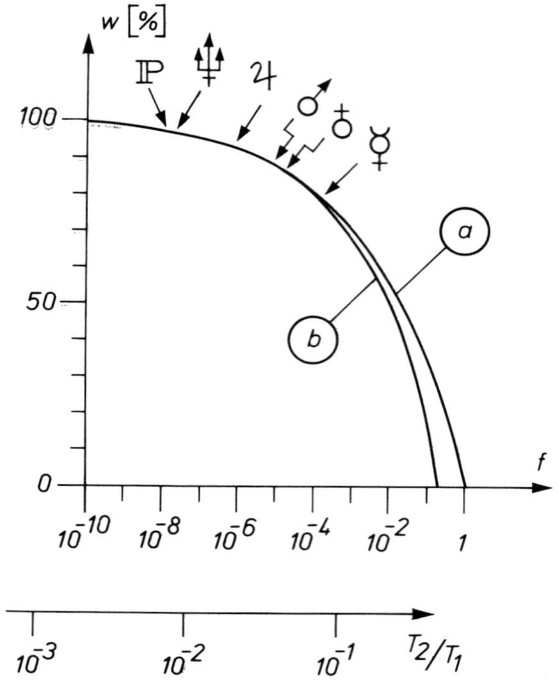

Fig. 5.6. Solar energy conversion efficiency w as a function of dilution factor f: (a) correct solution, (b) approximate solution.

The same result is obtained from eqn (5.13) by neglecting the second term of the left-hand side. Substituting the result into eqn (5.12) yields

$$w \approx 1 - \frac{5}{4}\, f^{1/5} .$$

Figure 5.6 shows the correct and the approximate solutions. If higher accuracy is required, one can prove the following series expansion [10] [13]:

$$w = 1 - \frac{5}{4}\, f^{1/5} + \frac{1}{4}\, f^{2/5} + \frac{9}{160}\, f^{3/5} + \dots .$$

Table 5.2 gives the results w, deduced from eqn (5.12), with η the solution of the exact equation (5.13), as well as the approximate w found by cutting off the series after the second term, for all planets except Venus.

Table 5.2. Maximum photothermal efficiency on the various planets.

Planet	f	T_p (K)	w	w_{approx}
Mercury	1.48×10^{-4}	449	0.793	0.786
Earth	2.16×10^{-5}	278	0.858	0.854
Mars	9.37×10^{-6}	225	0.879	0.877
Jupiter	8.01×10^{-7}	122	0.925	0.925
Saturn	2.38×10^{-7}	90	0.941	0.941
Uranus	5.89×10^{-8}	64	0.956	0.955
Neptune	2.40×10^{-8}	51	0.963	0.963
Pluto	1.44×10^{-8}	45	0.966	0.966

For the Earth, where

$$\frac{T_p}{T_s} = \frac{288 \text{ K}}{5762 \text{ K}} = 0.0500 \,,$$

we have that w equals 85.4 per cent. Note that the gain obtained by introducing light concentration is very substantial: $w = 86$ per cent instead of $w = 13$ per cent.

5.7 Selective black bodies

Engineers have found one more simple way of improving the Müser engine: by using a selective black absorber instead of a black one. This means that the top-side of the solar absorber has an emissivity $\epsilon(E)$ which is not equal to unity for all energies E, but in contrast obeys

$$\epsilon = \begin{cases} 0 & \text{for } E < E_g \\ 1 & \text{for } E > E_g \,. \end{cases} \tag{5.14}$$

The energy E_g is a material constant and is called the bandgap of the material. Thus the emitted radiation is zero for low energies and equals the Planck black-body spectrum for high energies. By applying the appropriate coating to a body, such a behaviour can be synthesized. Frequently applied coatings are metals (Be, Cr, Ni, W, ...) or metal oxides, fluorides or nitrides. Various layers can be combined. In order to demonstrate the performance of such coatings, Fig. 5.7(a) shows the experimental emissivity of a black body coated with an Al/SiO/Al/SiO-sandwich [7]. We recognize a bandgap E_g of 0.70 eV.

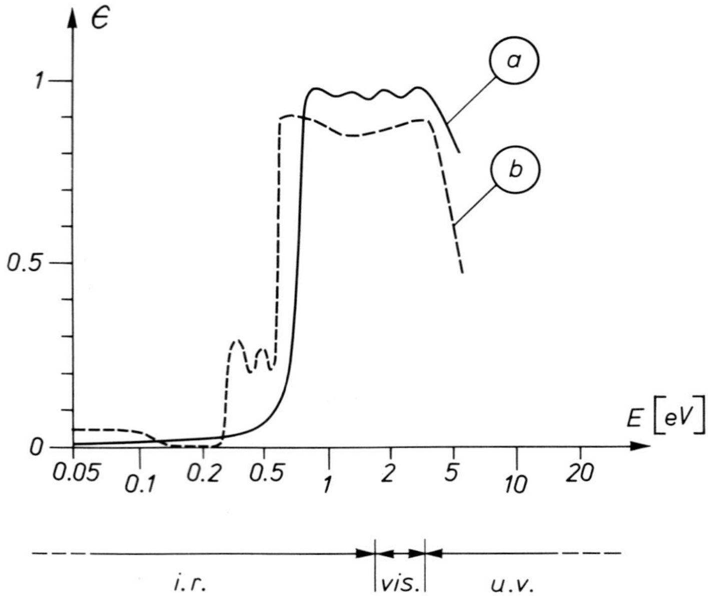

Fig. 5.7. Emissivity of a black body with cover: (a) with Al/SiO/Al/SiO coating, (b) with greenhouse window cover (i.r., infrared radiation; vis., visible light; u.v., ultraviolet radiation).

It is interesting to relate the above energy-dependent emissivity $\epsilon(E)$ with the greenhouse effect of the planets, as described in Chapter 2. Indeed, the emissivity of the planets was assumed to obey the following energy-dependency:

$$\epsilon = \begin{cases} 1 - \gamma & \text{for } E < E_g \\ 1 - \rho & \text{for } E > E_g \end{cases} ,$$

where E_g is an energy at the edge between the far infrared and the near infrared part of the spectrum.

In fact, the coatings described above are just a high-technology version of a greenhouse. Figure 5.7(b) shows the emissivity of a black body covered with an ordinary greenhouse window. We see that the modern coatings are just a way to approximate the function (5.14) better than the glass window does.

Why are bandgap materials so advantageous? For concentration values C which are not too high, the converter temperature T_3 is much lower than the Sun temperature T_s, such that the Planck spectrum of T_3 is

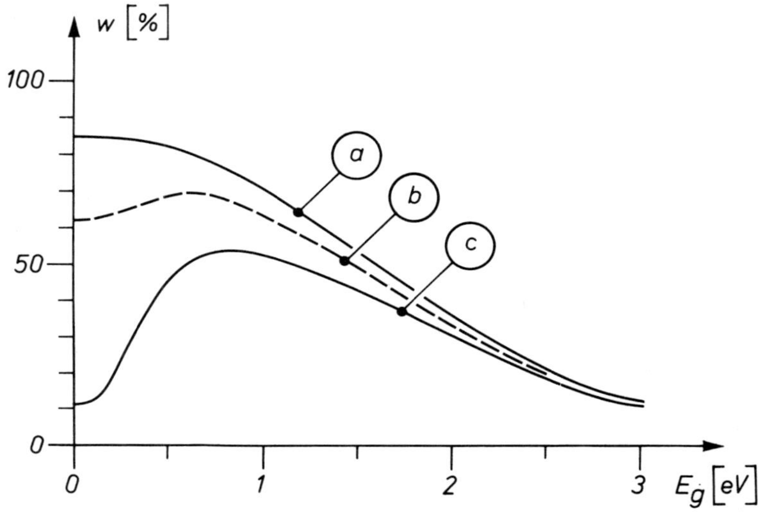

Fig. 5.8. Work produced by a photothermal solar energy convertor as a function of bandgap: (a) with fully concentrated sunlight (C = 46 300), (b) with moderately concentrated sunlight (C = 389), (c) with non-concentrated sunlight (C = 1).

concentrated at lower energies than the solar spectrum. By choosing ϵ equal to 0 at these low energies, we oppress the radiation emitted by the converter without lowering the solar radiation absorbed by the converter. As a consequence we can augment the net incoming radiation Q_1, as well as the resulting work W. This was extensively discussed by Castañs [5] and others [2] [3] [4] [6].

5.8 The Müser engine with bandgap

We now give Müser's solar absorber a bandgap. This means we use a selective coating, that only absorbs and emits photons having an energy E larger than the bandgap E_g of the coating material.

The introduction of the bandgap makes the constitutive law of the thermal conductor rather more complicated. By integrating Planck's spectrum from $E = E_g$ to $E = +\infty$, we obtain

$$Q_1 = g \, [\int_{E_g}^{\infty} \frac{E^3 dE}{\exp(\frac{E}{kT_1}) - 1} - \int_{E_g}^{\infty} \frac{E^3 dE}{\exp(\frac{E}{kT_3}) - 1}] \, , \tag{5.15}$$

instead of the simple Stefan–Boltzmann transport equation

$$Q_1 = g_1(T_1^4 - T_3^4) \; . \tag{5.16}$$

Note that the new equation, as well as the Stefan–Boltzmann rate equation, is of the general form

$$Q_1 = \text{function}(T_1) - \text{function}(T_3) \; .$$

The reader can easily verify that for $E_g \to 0$, eqn (5.15) becomes eqn (5.16), after defining $\frac{\pi^4 k^4}{15} g$ as g_1. For $E_g \neq 0$, eqn (5.15) cannot be simplified, as the integral

$$\int_a^{+\infty} \frac{x^3\, dx}{\exp(x) - 1}$$

cannot be evaluated analytically, if $a \neq 0$.

Written as a function of the fundamental parameter $\eta = 1 - T_2/T_3$, eqn (5.15) becomes

$$Q_1(\eta) = g\; [\int_{E_g}^{\infty} \frac{E^3 dE}{\exp(\frac{E}{kT_1}) - 1} - \int_{E_g}^{\infty} \frac{E^3 dE}{\exp(\frac{E}{kT_2}(1-\eta)) - 1}\;] \; .$$

The corresponding curves $Q_1(\eta)$, $W(\eta)$ and $\Sigma S(\eta)$ are very similar to those of the Müser engine with zero bandgap (Fig. 3.8). Again, maximum-power values of η, Q_1, W, and ΣS have to be calculated numerically. Again, $Q_1 = 0$, $W = 0$ and $\Sigma S = 0$ at $\eta = \eta_C$.

With our two-source model (the Sun at temperature T_s and the surroundings at temperature T_p), the equation for Q_1 becomes

$$Q_1 = g\; [\; Cf \int_{E_g}^{\infty} \frac{E^3 dE}{\exp(\frac{E}{kT_s}) - 1} \; + \; (1 - Cf) \int_{E_g}^{\infty} \frac{E^3 dE}{\exp(\frac{E}{kT_p}) - 1} - \int_{E_g}^{\infty} \frac{E^3 dE}{\exp(\frac{E}{kT_p}(1-\eta)) - 1}\;] \; . \tag{5.17}$$

Figure 5.8 shows numerical results for $f = 2.16 \times 10^{-5}$, i.e. for the Earth, and for $C = 1$, $C = 389$, and $C = 46\,300$, i.e. in the absence of concentration, for intermediate concentration and for the highest concentration physically possible. We see that for $Cf = 1$ the optimal bandgap is zero: the best converter is the Müser engine without bandgap. But for any $C < 1/f$, use of a bandgap coating is advantageous. For example, for natural sunlight, i.e. for $C = 1$, the optimal bandgap is 0.90 eV, which gives rise to an efficiency of 53.6 per cent.

Figure 5.9 shows the optimal-bandgap solar energy efficiencies for different temperature ratios T_p/T_s. The result for the Earth, with natural sunlight, i.e. $w = 53.6$ per cent, can be seen on curve (c), for $T_p/T_s = 0.0500$.

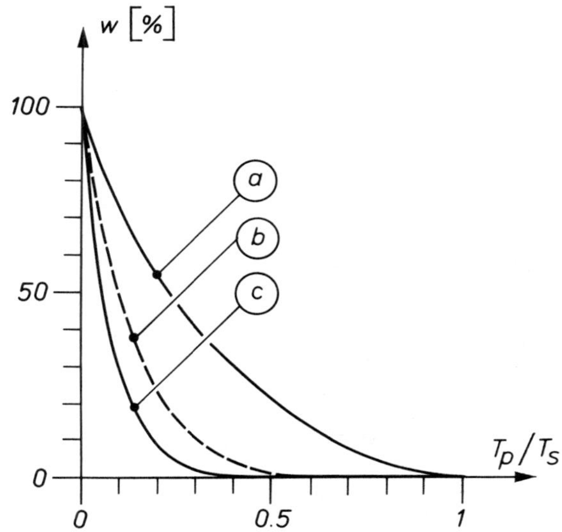

Fig. 5.9. Work produced by a photothermal solar energy convertor as a function of temperature ratio T_p/T_s : (a) with fully concentrated sunlight ($C = 46\,300$), (b) with moderately concentrated sunlight ($C = 389$), (c) with non-concentrated sunlight ($C = 1$).

5.9 Conclusion

We have seen in this chapter that the simple Müser engine can convert the solar energy with an efficiency of 12.8 per cent. There are, however, two ways to do better:

- By introducing a bandgap, the performance can be improved substantially. The actual gain depends upon the f, i.e. upon our position in the solar system. For the Earth, we can achieve 53.6 per cent.
- By introducing light concentration, we can still do better. A high efficiency, slightly dependent on f, can be reached. For the Earth, we can get 85.4 per cent.

5.10 An example

As an example of a photothermal converter, we consider the solar farm 'Eurelios' at Adrano (Italy). An array of 182 flat mirrors (with a total

Fig. 5.10. Solar power plant 'Eurelios'.

surface area of 6216 m^2) reflects the solar rays towards an absorbing cavity (with an aperture area of 16 m^2), on top of a tower 55 m high. The heat, collected in this solar panel, is converted into electrical power by means of a turbine and an alternator. See Fig. 5.10.

The concentration factor amounts to

$$C = \frac{6216 \text{ m}^2}{16 \text{ m}^2} = 389 \ .$$

From Fig. 5.8(b) we deduce that the optimal cavity coating has a bandgap of 0.58 eV, realizing a conversion efficiency of 68.2 per cent. The experimental efficiency [12] amounts to

$$w = \frac{750 \text{ kW}}{5.7 \text{ MW}} = 0.13 \ .$$

The large discrepancy between the experimental value of 13 per cent and the upper bound of 68 per cent is explained by the fact that real installations

- have mirrors with far from ideal reflectivity,

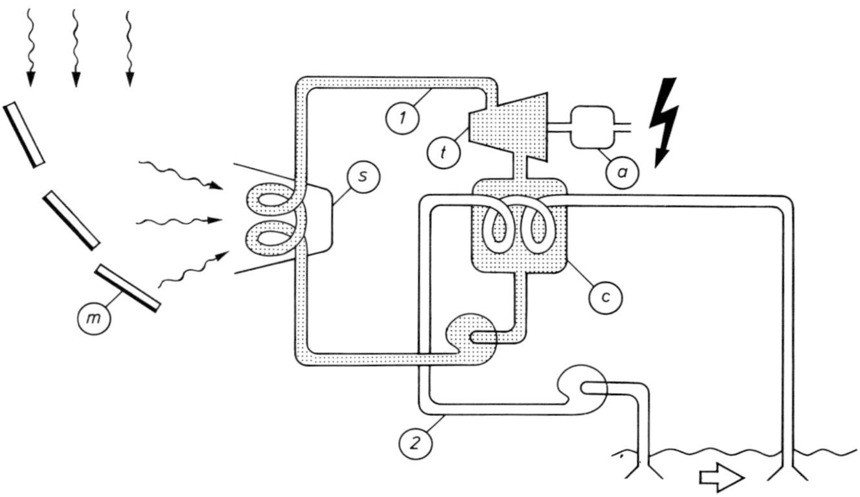

Fig. 5.11. Solar power plant: (1) primary circuit, (2) secondary circuit, (m) mirrors, (s) steam generator, (t) turbine, (a) alternator, (c) condensor.

- have an absorber coating with far from ideal optical selectivity $\epsilon(E)$, as good refractory properties are the design priority.

Further, the engine between reservoir 3 and reservoir 2 is not a reversible Carnot engine, but rather an engine that resembles well an endoreversible engine with linear conductances g_1 (steam generator) and g_2 (condensor): see Fig. 5.11. Thus a real solar power tower is a combination of a Müser engine and a Curzon–Ahlborn engine: see Fig. 5.12.

Bibliography

1. H. Müser (1957). Behandlung von Elektronenprozessen in Halbleiter-Randschichten. *Zeitschrift für Physik*, volume 148, pp. 380 - 390.

2. L. Shaffer (1958). Wavelength-dependent (selective) processes for the utilization of solar energy. *Solar Energy*, volume 2, pp. 21 - 26.

3. C. Liebert and R. Hibbard (1962). Spectrally selective collector surfaces in a solar-driven Carnot space-power system. *Solar Energy*, volume 6, pp. 84 - 88.

4. R. Cross (1963). Collector surfaces of solar-powered thermal systems. *Solar Energy*, volume 7, pp. 152 - 155.

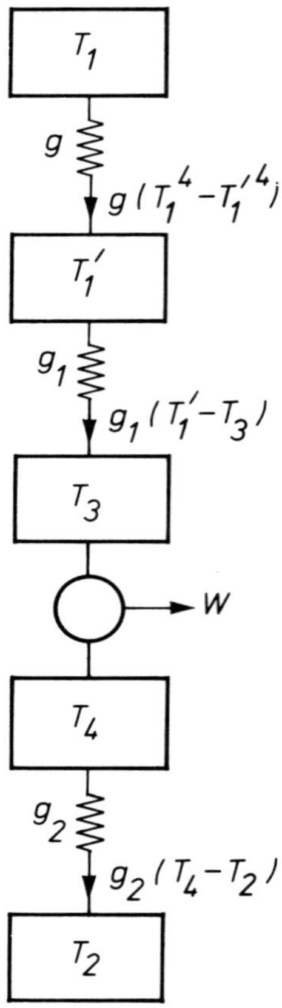

Fig. 5.12. Realistic solar engine.

5. M. Castañs (1976). Bases físicas del approvechamiento de la energía solar. *Revista Geofísica*, volume 35, pp. 227 - 239.

6. J. Williams (1977). *Solar energy: technology and applications*. Ann Arbor Science Publishers, Ann Arbor, pp. 17 - 37 and 87 - 97.

7. W. Palz (1978). *Solar electricity*. Butterworths-Unesco, Paris, pp. 73 - 83 and 157- 177.

8. S. Jeter (1981). Maximum conversion efficiency for the utilization of direct solar radiation. *Solar Energy*, volume 26, pp. 231 - 236.

9. A. De Vos and H. Pauwels (1981). On the thermodynamic limit of photovoltaic solar energy conversion. *Applied Physics*, volume 25, pp. 119 - 125.

10. A. De Vos and H. Pauwels (1983). Comment on a thermodynamic paradox presented by P. Würfel. *Journal of Physics C: Solid-State Physics*, volume 16, pp. 6897 - 6909.

11. C. Etievant (1984). Solar central receiver power plants. *International Journal of Solar Energy*, volume 2, pp. 309 - 339.

12. J. Gretz (1987). Concept and operating experiences with Eurelios 1 MW helioelectric power plant of the European Communities. *I.E.E. Proceedings~A: Science, Measurements and Technology* , volume 134, pp. 399 - 406.

13. A. De Vos (1989). The use of `Reduce` in solar energy conversion theory. *Reports of the* `CAGe` *Project*, volume 4, pp. 1 - 17.

6
Photovoltaic conversion

6.1 Introduction

Photothermal conversion of solar energy into work is sometimes referred to as an indirect conversion, i.e. conversion of solar energy into heat by collection in the solar absorber and subsequent conversion of this heat into work by means of the Carnot engine. In this chapter we will discuss the alternative approach of solar energy conversion: direct conversion, called photovoltaic conversion.

6.2 Semiconductors

Like the selective black bodies of Chapter 5, the semiconductors applied in this chapter are bandgap materials. Electron levels exist only in two energy bands, called valence band and conduction band. See Fig. 6.1. No intraband radiative interactions are possible: only interband radiative transitions are possible. Thus only photons with energy E larger than the bandgap E_g can be absorbed or emitted.

What makes semiconductors special is as follows. All electron levels of the valence band are in thermodynamic equilibrium: they have the same temperature T and the same Fermi energy E_F. All electron levels of the conduction band are in thermal equilibrium with each other: they have the same temperature T and the same Fermi energy E_F. But the levels of the valence band are not completely in thermodynamic equilibrium with the levels of the conduction band: both bands have the same temperature T, but different Fermi energies E_{Fv} and E_{Fc}. According to the theory of Sections 1.2 and 1.3, such a material has the following emission spectrum:

$$n(E) = \begin{cases} 0 & \text{for } E < E_g \\ \dfrac{2\pi}{c^2h^3} \dfrac{E^2}{\exp(\dfrac{E-\mu}{kT}) - 1} & \text{for } E > E_g \end{cases} ,$$

where the chemical potential μ of the photons is given by

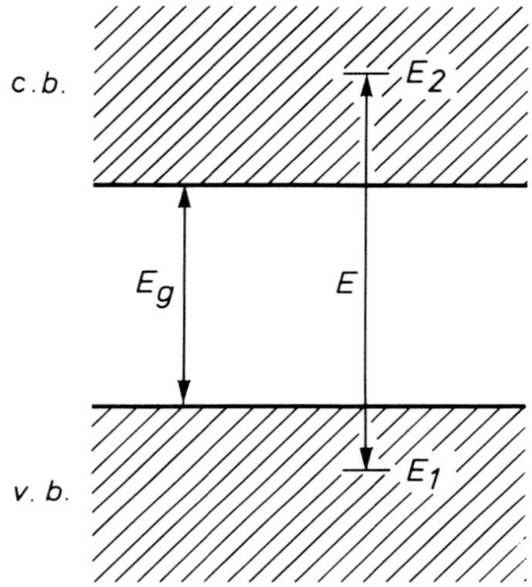

Fig. 6.1. Semiconductor band diagram (c.b., conduction band; v.b., valence band).

$$\mu = E_{Fc} - E_{Fv} \ .$$

In the above equation, we tacitly assumed a unitary emissivity ϵ for all $E > E_g$.

With a semiconductor, we can make an electronic device called diode, by applying two ohmic contacts. This is realized by means of two metal wires. Metals are materials with no bandgap: all electrons in the same metal wire have the same Fermi energy. One wire has its Fermi energy equal to E_{Fv} (the ohmic contact to the valence band) and one wire has its Fermi energy equal to E_{Fc} (the ohmic contact to the conduction band).

Between two metal wires we can apply an electric voltage V:

$$V = \frac{E_{Fc} - E_{Fv}}{q} \ ,$$

where q is the elementary charge introduced at the end of Section 1.4.

Therefore, the two ohmic contacts allow us to change externally or bias the Fermi gap $E_{Fc} - E_{Fv}$ and thus the chemical potential of the light:

$$\mu = qV \ .$$

We now consider the photon exchange between two biased semiconductors facing each other. The net energy flux U and the net particle flux N are given by

$$\begin{aligned} U &= g\,[\int_{E_g}^{\infty} \frac{E^3 dE}{\exp(\frac{E-\mu_1}{kT_1})-1} - \int_{E_g}^{\infty} \frac{E^3 dE}{\exp(\frac{E-\mu_2}{kT_2})-1}\,] \\ N &= g\,[\int_{E_g}^{\infty} \frac{E^2 dE}{\exp(\frac{E-\mu_1}{kT_1})-1} - \int_{E_g}^{\infty} \frac{E^2 dE}{\exp(\frac{E-\mu_2}{kT_2})-1}\,]\,, \end{aligned} \tag{6.1}$$

where

$$\begin{aligned} g &= \frac{2\pi}{c^2 h^3} \\ &= \frac{15}{\pi^4 k^4}\,\sigma\,. \end{aligned}$$

These expressions are of the form

$$\begin{aligned} U &= \text{function}_1(T_1,\mu_1) - \text{function}_1(T_2,\mu_2) \\ N &= \text{function}_2(T_1,\mu_1) - \text{function}_2(T_2,\mu_2)\,. \end{aligned} \tag{6.2}$$

This constitutes a generalization of the form

$$\begin{aligned} U &= \text{function}_1(T_1) - \text{function}_1(T_2) \\ N &= \text{function}_2(T_1) - \text{function}_2(T_2)\,, \end{aligned}$$

which we have encountered until now. For example, we had for two black bodies

$$\begin{aligned} U &= \sigma(T_1^4 - T_2^4) \\ N &= \sigma'(T_1^3 - T_2^3)\,, \end{aligned}$$

whereas for selective black bodies we had

$$\begin{aligned} U &= g\,[\int_{E_g}^{\infty} \frac{E^3 dE}{\exp(\frac{E}{kT_1})-1} - \int_{E_g}^{\infty} \frac{E^3 dE}{\exp(\frac{E}{kT_2})-1}\,] \\ N &= g\,[\int_{E_g}^{\infty} \frac{E^2 dE}{\exp(\frac{E}{kT_1})-1} - \int_{E_g}^{\infty} \frac{E^2 dE}{\exp(\frac{E}{kT_2})-1}\,]\,. \end{aligned}$$

For these two examples thermodynamic equilibrium between reservoir 1 and reservoir 2 is fulfilled as soon as $T_1 = T_2$. For our new semiconductor model, thermodynamic equilibrium requires two conditions to be fulfilled: $T_1 = T_2$ and $\mu_1 = \mu_2$.

Figure 6.2 shows the two diodes facing each other. It is interesting to note that such a configuration is actually applied in telecommunication, where it is called a photocoupler.

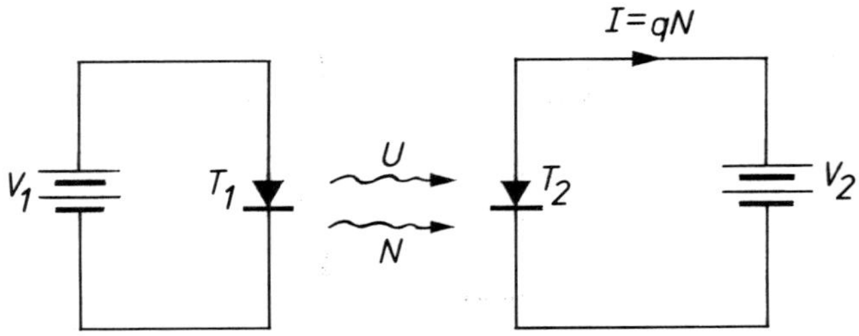

Fig. 6.2. Photocoupler.

Instead of letting reservoir 1 and reservoir 2 radiate freely to one another, we now insert a thermodynamic engine in between: see Fig. 6.3(a). Note that in the thermodynamic diagram the two reservoirs are labelled by their constant temperature T and their constant chemical potential μ. The diagram is therefore an extention of the Carnot diagram of Fig. 3.1(a), where reservoirs were labelled only by a temperature. The endoreversible extension of Fig. 6.3(a) is given by Fig. 6.3(b), where we have four reservoirs.

In solar energy conversion, there is of course no diode 1 and no diode 2. Neither the source 1 nor the sink 2 can be biased by any voltage V_1 or V_2. The source is the Sun, i.e. a black body. Thus $\mu_1 = 0$. The sink is our planetary environment, i.e. also a black body. Thus $\mu_2 = 0$. Therefore only μ_3 and μ_4 can be different from zero. Therefore, we finally obtain the configuration of Fig. 6.3(c).

6.3 The solar cell

In photovoltaic solar energy conversion, the converter, i.e. the solar cell, is assumed to be in perfect thermal contact with the surroundings, i.e. the planet: $T_3 = T_4 = T_2 = T_p$. See Fig. 6.3(d).

In an ideal solar cell, each photon absorbed causes an electron to jump from the valence band to the conduction band and thus leads to an electric current pulse equal to q, the elementary charge. Equally, each electron falling from the conduction band to the valence band gives rise to the emission of a photon, as well as to an electric current pulse $-q$. In such an idealized cell any non-radiative decay from conduction band to valence band is absent. The electric current I through this cell is given by

$$I = qN ,$$

where N is the net particle flux, absorbed by the cell.

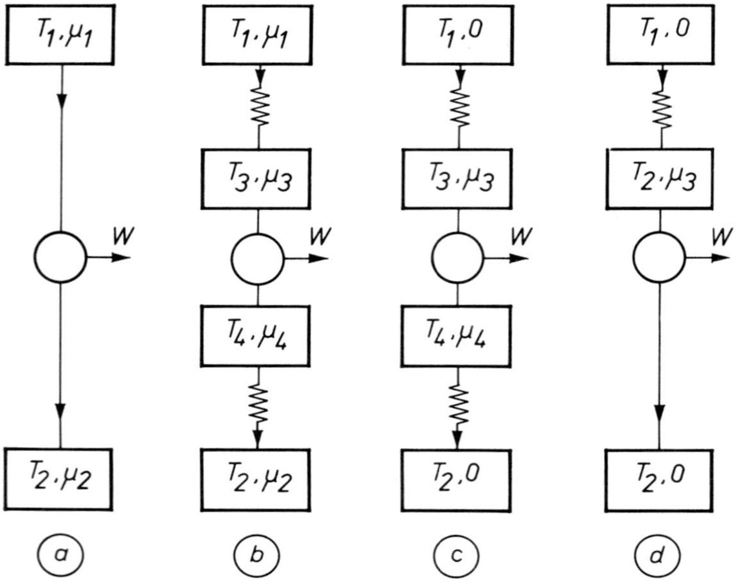

Fig. 6.3. Four thermodynamic configurations: (a) reversible engine, (b) endoreversible engine, (c) solar engine, (d) solar cell.

Thus the $I(V)$ characteristic or current-voltage characteristic of a solar cell is given by

$$I(V) = qg \left[Cf \int_{E_g}^{\infty} \frac{E^2 dE}{\exp(\frac{E}{kT_s}) - 1} + (1 - Cf) \int_{E_g}^{\infty} \frac{E^2 dE}{\exp(\frac{E}{kT_p}) - 1} - \int_{E_g}^{\infty} \frac{E^2 dE}{\exp(\frac{E - qV}{kT_p}) - 1} \right] . \quad (6.3)$$

This equation is very similar to the $Q_1(\eta)$ characteristic (5.17) of the heat engine discussed in Chapter 5. Just like the Müser characteristic (and any Stefan–Boltzmann engine characteristic), the solar cell characteristic displays two asymptotes and two particular points:

- For $V = -\infty$, we have the saturation current:

$$\begin{aligned} I_{sat} &= I(-\infty) \\ &= qg \left[Cf \int_{E_g}^{\infty} \frac{E^2 dE}{\exp(\frac{E}{kT_s}) - 1} + (1 - Cf) \int_{E_g}^{\infty} \frac{E^2 dE}{\exp(\frac{E}{kT_p}) - 1} \right] . \end{aligned}$$

- For $V = 0$, we have the short-circuit current:

$$I_{sc} = I(0) = qgCf\ \left[\int_{E_g}^{\infty} \frac{E^2 dE}{\exp(\frac{E}{kT_s}) - 1} - \int_{E_g}^{\infty} \frac{E^2 dE}{\exp(\frac{E}{kT_p}) - 1} \right] .$$

- For $I = 0$, we have an open-circuit voltage, defined by

$$I(V_{oc}) = 0 \ .$$

- For $V = E_g/q$, the current becomes infinite:

$$I(E_g/q) = -\infty \ .$$

In Sections 3.3 and 3.6, the short-circuit point and the open-circuit point divided the $Q_1(\eta)$ characteristic of a thermal engine into three distinct parts. Analogously, the short-circuit and open-circuit points of a solar cell divide the $I(V)$ into three parts:

- For $-\infty < V < 0$, the diode works as a photodiode.
- For $0 < V < V_{oc}$, the diode works as a true solar cell.
- For $V_{oc} < V < E_g/q$, the diode works as a light-emitting diode (LED).

It is no surprise that we encounter here a measuring device: the photodiode. The latter is the electronic equivalent of the thermographic camera, extensively described in Section 3.8. Whereas the cooling of the camera (i.e. $T_3 < T_2$) guaranteed precise measurement of the heat flow emitted by reservoir 1, the negative biasing of the diode (i.e. $\mu_3 < \mu_2 = 0$) guarantees precise measurement of the particle flow emitted by reservoir 1.

Whereas in the photothermal engine $\lim_{\eta \to 1} Q_1 = -\infty$, we now have in the photovoltaic engine $\lim_{V \to E_g/q} I = -\infty$. It is therefore no surprise that in photovoltaics the dimensionless number $\zeta = qV/E_g$ plays the same role as η in photothermal conversion.

We now remark that the electrical power $W = VI$ can be written as $W = \zeta E_g N(\zeta)$, where the function $N(\zeta)$ is given by

$$N(\zeta) = g\ \left[Cf \int_{E_g}^{\infty} \frac{E^2 dE}{\exp(\frac{E}{kT_s}) - 1} + (1 - Cf) \int_{E_g}^{\infty} \frac{E^2 dE}{\exp(\frac{E}{kT_p}) - 1} - \int_{E_g}^{\infty} \frac{E^2 dE}{\exp(\frac{E - \zeta E_g}{kT_p}) - 1} \right] .$$

For the energy exchange with the Sun, we have:

$$U(\zeta) = g \, [\, Cf \int_{E_g}^{\infty} \frac{E^3 dE}{\exp(\frac{E}{kT_s}) - 1} \; + \; (1 - Cf) \int_{E_g}^{\infty} \frac{E^3 dE}{\exp(\frac{E}{kT_p}) - 1} \; - \; \int_{E_g}^{\infty} \frac{E^3 dE}{\exp(\frac{E - \zeta E_g}{kT_p}) - 1} \,] \, .$$

The corresponding curves $N(\zeta)$, $U(\zeta)$, and $W(\zeta)$ are shown in Fig. 6.4. The figure also displays the entropy creation rate $\Sigma S(\zeta)$.

Numerical calculations lead to the solar energy efficiency

$$w = VI(V)/Cf\sigma T_s^4$$

of Fig. 6.5. If we compare these results with Fig. 5.8, we can see two important differences:

- the efficiency of photovoltaic conversion is much lower than the efficiency of photothermal coversion, and
- for any C the optimal bandgap is positive, whereas maximum C gave rise to zero optimal bandgap in the photothermal case.

For the f value of the Earth we have the following numerical results:

- for unconcentrated light (i.e. $C = 1$) we have $w = 31.0$ per cent at the optimal bandgap $E_g = 1.30$ eV.
- for fully concentrated light (i.e. $C = 1/f$) we have $w = 40.8$ per cent at the optimal bandgap $E_g = 1.15$ eV.

For other f values, Fig. 6.6 shows the maximum efficiency (i.e. efficiency at optimal bandgap) as a function of the temperature ratio T_p/T_s. If we compare this figure with Fig. 5.9, again the low efficiency of photovoltaics is striking. Whereas for photothermal conversion we had $\lim_{T_p \to 0} w = 1$, now the limit of w for T_p (and thus f) going to zero is only 44 per cent.

Let us conclude this section by trying to calculate an analytical expression for this upper limit of 44 per cent. We perform the calculations for extreme light concentration, i.e. for $Cf = 1$. We have

$$W = Vqg \, [\int_{E_g}^{\infty} \frac{E^2 dE}{\exp(\frac{E}{kT_s}) - 1} - \int_{E_g}^{\infty} \frac{E^2 dE}{\exp(\frac{E - qV}{kT_p}) - 1} \,] \, . \qquad (6.4)$$

We consider W as a function of two variables: the bandgap E_g and the voltage V.

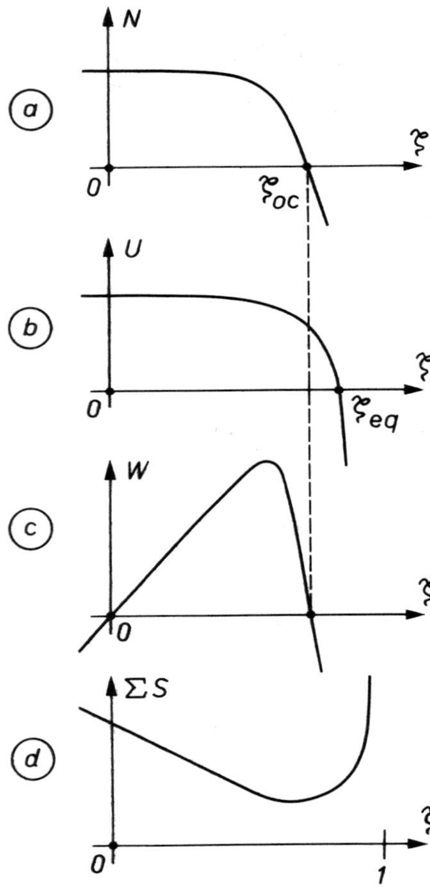

Fig. 6.4. Thermodynamic quantities as a function of dimensionless electrical voltage $\zeta = qV/E_g$ of a solar cell: (a) particle consumption N, (b) energy consumption U, (c) work production W, (d) entropy creation rate ΣS.

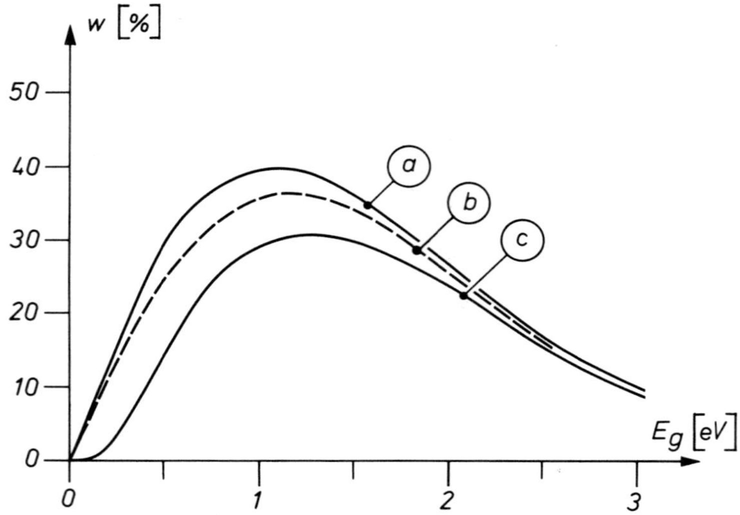

Fig. 6.5. Work produced by a photovoltaic solar energy converter as a function of bandgap: (a) with fully concentrated sunlight (C = 46 300), (b) with moderately concentrated sunlight (C = 389), (c) with non-concentrated sunlight (C = 1).

First we express that we want an optimal bandgap:

$$\frac{\partial W}{\partial E_g} = 0 \ .$$

Taking into account the simple rule of mathematics

$$\frac{\partial}{\partial a} \int_a^b f(x)dx = -f(a) \ ,$$

we find

$$qV = E_g(1 - \frac{T_p}{T_s}) \ .$$

Substitution of this result into eqn (6.4) yields

$$W = gE_g(1-\frac{T_p}{T_s}) \ [\int_{E_g}^{\infty} \frac{E^2dE}{\exp(\frac{E}{kT_s}) - 1} - \int_{E_g}^{\infty} \frac{E^2dE}{\exp(\frac{E}{kT_p} - \frac{E_g}{kT_p} + \frac{E_g}{kT_s}) - 1}] \ .$$

By introducing a new integration variable

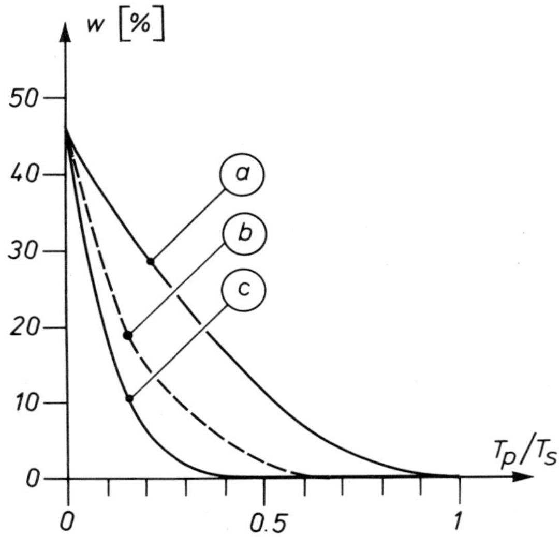

Fig. 6.6. Work produced by a photovoltaic solar energy converter as a function of temperature ratio T_p/T_s: (a) with fully concentrated sunlight ($C = 46\,300$), (b) with moderately concentrated sunlight ($C = 389$), (c) with non-concentrated sunlight ($C = 1$).

$$E' = \frac{T_s}{T_p}E - \frac{T_s}{T_p}E_g + E_g$$

in the second integral, we get

$$W = gE_g(1 - \frac{T_p}{T_s}) \quad [\quad \int_{E_g}^{\infty} \frac{E^2 dE}{\exp(\frac{E}{kT_s}) - 1} - \frac{T_p}{T_s} \int_{E_g}^{\infty} \frac{(\frac{T_p}{T_s}E' + E_g - \frac{T_p}{T_s}E_g)^2 dE'}{\exp(\frac{E'}{kT_s}) - 1} \;] \, .$$

For the special case $T_p \to 0$, this finally yields

$$W = gE_g \int_{E_g}^{\infty} \frac{E^2 dE}{\exp(\frac{E}{kT_s}) - 1} \, .$$

By introducing the dimensionless bandgap $x_g = E_g/kT_s$ and after division by σT_s^4, we have

$$w = \frac{15}{\pi^4}\, x_g \int_{x_g}^{\infty} \frac{x^2 dx}{\exp(x) - 1} \; . \tag{6.5}$$

Maximization of w with respect to x_g yields the condition

$$\int_{x_g}^{\infty} \frac{x^2 dx}{\exp(x) - 1} - \frac{x_g^3}{\exp(x_g) - 1} = 0 \; .$$

This transcendental equation has the solution $x_g = 2.166$. Substitution of this value in eqn (6.5) yields

$$w = 0.440 \; .$$

This 44 per cent fundamental upper limit of single diode solar cells was discovered by Trivich and Flinn [1] in 1955 and rediscovered by Shockley and Queisser [2] in 1961.

We can ask ourselves: why exactly 44 per cent? The reader will easily guess that this number is related to the dimensionality of our space. And indeed, Table 6.1 gives the fundamental upper limit of the n-dimensional solar cell [11]:

$$\begin{aligned} w &= \frac{x_g \int_{x_g}^{\infty} \frac{x^{n-1} dx}{\exp(x) - 1}}{\int_0^{\infty} \frac{x^n \, dx}{\exp(x) - 1}} \\ &= \frac{1}{\Gamma(n+1)\zeta(n+1)}\, x_g \int_{x_g}^{\infty} \frac{x^{n-1} dx}{\exp(x) - 1} \; , \end{aligned}$$

where x_g is the solution of

$$\int_{x_g}^{\infty} \frac{x^{n-1} dx}{\exp(x) - 1} - \frac{x_g^n}{\exp(x_g) - 1} = 0 \; .$$

We see how the upper bound of w increases monotonically with n. For large n, it increases as $1 - \sqrt{\log(n)/n}$.

This will constitute our last journey through higher dimensional space. From now on, we will wisely stay in the three-dimensional world.

6.4 An example

An example of a medium-sized photovoltaic plant is the autonomous photovoltaic power system of the remote fisher village *Αγια Ρουμελη* or Aghià Roumeli (Greece). The 720 modules (each containing 68 solar cells 100 mm in diameter) have a total surface area of 596 m^2. See Fig. 6.7.

Table 6.1. Upper limit of photovoltaic solar energy conversion.

n	w
1	0.292
2	0.385
3	0.440
4	0.476
5	0.504
6	0.525
7	0.543
8	0.558
9	0.571
10	0.582
11	0.593
12	0.602
...	
25	0.679
...	

In summer time, outside temperature is approximately 30 °C. The solar cells are made of silicon, a 1.12 eV bandgap semiconductor. As no concentration optics is applied, we have $C = 1$. From Fig. 6.5 we see that $E_g =$ 1.12 eV is only slightly smaller than optimum (E_g = 1.30 eV) and gives rise to w = 30.9 per cent. The experimental value [7] is

$$w = \frac{50 \text{ kW}}{596 \text{ kW}} = 0.084 \ .$$

The large discrepancy between actual and theoretically possible efficiency can partly be attributed to the fact that the round silicon wafers do not cover the panel area completely. Indeed, we can easily calculate that the cells have a surface area of only 385 m^2, i.e. only 65 per cent of the total panel area. If we calculate the solar energy conversion efficiency, taking into account only the solar cell area of the photovoltaic panels, we get

$$w = \frac{50 \text{ kW}}{385 \text{ kW}} = 0.13 \ .$$

This number is still much smaller than the theoretical limit of 31 per cent. This is mainly caused by the fact that real semiconductors

- do not succeed in collecting all incident photons with energy larger than the bandgap, and

Fig. 6.7. Aghià Roumeli photovoltaic power plant.

- suffer from non-radiative decay processes besides the two radiative ones of Section 1.3 (spontaneous emission of a photon and stimulated emission of a photon).

Systematic increase in silicon quality, as well as cell and panel design, is illustrated in Fig. 6.8, which displays the photovoltaic conversion efficiencies of both laboratory silicon solar cells and commercially available silicon solar cell panels [8] [9] [10].

Bibliography

1. D. Trivich and P. Flinn (1955). Maximum efficiency of solar energy conversion by quantum processes. In *Solar energy research* (eds. F. Daniels and J. Duffie). Thames and Hudson, London, pp. 143 - 147.

2. W. Shockley and H. Queisser (1961). Detailed balance limit of efficiency of p-n junction solar cells. *Journal of Applied Physics*, volume 32, pp. 510 - 519.

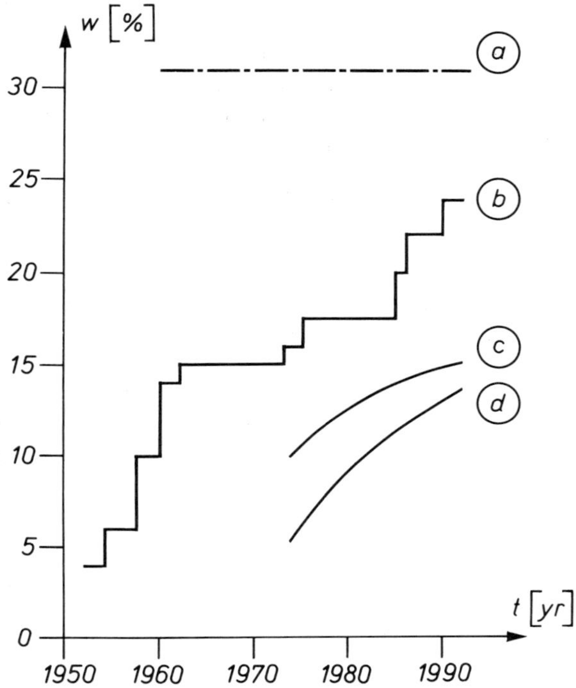

Fig. 6.8. Progress in terrestrial silicon photovoltaics ($C = 1$) : (a) thermodynamic limit, (b) highest cell efficiency ever realized in the laboratory, (c) cell efficiency in industrial production, (d) module efficiency in industrial production.

3. W. Palz (1978). *Solar electricity.* Butterworths-Unesco, Paris, pp. 179 - 223.

4. A. De Vos and H. Pauwels (1981). On the thermodynamic limit of photovoltaic solar energy conversion. *Applied Physics*, volume 25, pp. 119 - 125.

5. H. Pauwels and A. De Vos (1981). Determination and thermodynamics of the maximum efficiency photovoltaic device. *Proceedings of the 15 th I.E.E.E. Photovoltaic Specialists Conference*, Orlando, 11 - 15 May 1981, pp. 377 - 382.

6. C. Kourogenis, B. Aubert, L. Selles and J. Chadjivassiliadis (1981). Aghià Roumeli - island of Creta, a 50 kW autonomous photovoltaic

power. In *Medium-size photovoltaic power plants* (eds. H. Durand, P. Maycock and W. Palz). Reidel, Dordrecht, pp. 121 -123.

7. B. Aubert (1983). Aghià Roumeli electricity supply to an isolated village. In *Solar energy research and development in the European Community, Series C: photovoltaic power generation, volume 4* (ed. W. Paltz). Reidel, Dordrecht, pp. 187 - 201.

8. E. Christensen (1985). *Flat-plate solar array project of the U.S. Department of Energy's national photovoltaics program: 10 years of progress.* Jet Propulsion Laboratory, report no. JPL 400-279 10/85.

9. M. Green (1987). High efficiency silicon solar cells. Trans Tech Publications, Aedermannsdorf, pp. 139 - 168.

10. W. Bloss and F. Pfisterer (1990). A general view on photovoltaic cells. *Proceedings of the 8 th General Conference of the European Physical Society*, Amsterdam, 4 - 8 September 1990, pp. 611 - 630.

11. P. T. Landsberg, A. De Vos and P. Baruch (1991). A comparison of some efficiency factors in photovoltaics. *Journal of Physics: Condensed Matter*, volume 3, pp. 6415 - 6424.

7 Hybrid conversion

7.1 Introduction

In this chapter, we will compare photovoltaic conversion to photothermal conversion, in the framework of endoreversible thermodynamics. First we will demonstrate that photovoltaics is somewhat more complicated than photothermal energy conversion. We will then show that both can be presented as special cases of a more general endoreversible engine: the so-called hybrid pt/pv (photothermal/photovoltaic) converter.

7.2 Is a solar cell really an endoreversible engine? ♡

Let us consider a solar cell at temperature T_2, facing a light source at temperature T_1. For convenience, we assume that the source and the cell 'surround' each other completely: all photons leaving the source are incident on the cell and vice versa. This corresponds with ultimate concentration, i.e. $C = 1/f$.

Under this condition, the net absorbed particle flux is

$$N(V) = g\ [\int_{E_g}^{\infty} \frac{E^2 dE}{\exp(\frac{E}{kT_1}) - 1} - \int_{E_g}^{\infty} \frac{E^2 dE}{\exp(\frac{E-qV}{kT_2}) - 1}\ ]\ ,$$

where V is the voltage over the solar cell. Using the dimensionless voltage $\zeta = qV/E_g$, this becomes

$$N(\zeta) = g\ [\int_{E_g}^{\infty} \frac{E^2 dE}{\exp(\frac{E}{kT_1}) - 1} - \int_{E_g}^{\infty} \frac{E^2 dE}{\exp(\frac{E-\zeta E_g}{kT_2}) - 1}\ ]\ .$$

The curve $N(\zeta)$ on Fig. 6.4(a) displays an open-circuit point at $\zeta = \zeta_{oc}$, where ζ_{oc} is given by the equation

$$\int_{E_g}^{\infty} \frac{E^2 dE}{\exp(\frac{E}{kT_1}) - 1} - \int_{E_g}^{\infty} \frac{E^2 dE}{\exp(\frac{E-\zeta_{oc} E_g}{kT_2}) - 1} = 0\ .$$

This expression cannot be simplified, and therefore cannot be solved analytically for ζ_{oc}. We can, however, prove that

$$\eta_C < \zeta_{oc} < 1 \ .$$

An alarming feature appears now. From Fig. 6.4(b) we see that, at this particular value ζ_{oc} of ζ, the energy exchange $U(\zeta_{oc})$ is not zero! Indeed, $U(\zeta)$ becomes zero for another value of ζ, say ζ_{eq}. This 'equilibrium' value of ζ is given by the equation

$$\int_{E_g}^{\infty} \frac{E^3 dE}{\exp(\frac{E}{kT_1}) - 1} - \int_{E_g}^{\infty} \frac{E^3 dE}{\exp(\frac{E-\zeta_{eq}E_g}{kT_2}) - 1} = 0 \ .$$

Neither this equation can be simplified. It thus also has to be solved numerically. One finds that

$$\eta_C < \zeta_{oc} < \zeta_{eq} < 1 \ .$$

In the photothermal case, open-circuit condition $\eta_{oc} = 1 - T_2/T_1$ not only implied balance of the incoming energy with the outgoing energy:

$$\int_{E_g}^{\infty} \frac{E^3 dE}{\exp(\frac{E}{kT_1}) - 1} = \int_{E_g}^{\infty} \frac{E^3 dE}{\exp(\frac{E}{kT_2}(1-\eta_{oc})) - 1} \ ,$$

guaranteeing $U = 0$, but we had more: detailed balance. Absorbed and emitted radiation were in balance at every frequency interval of the two spectra:

$$\frac{E^2 dE}{\exp(\frac{E}{kT_1}) - 1} = \frac{E^2 dE}{\exp(\frac{E}{kT_2}(1-\eta_{oc})) - 1} \ ,$$

so that automatically

$$\int_{E_g}^{\infty} \frac{E^2 dE}{\exp(\frac{E}{kT_1}) - 1} = \int_{E_g}^{\infty} \frac{E^2 dE}{\exp(\frac{E}{kT_2}(1-\eta_{oc})) - 1}$$

was also fulfilled and hence $N = 0$. Such a coincidence of $U = 0$ and $N = 0$ is not the case in the present photovoltaic conversion. Under open-circuit conditions, we have here an overall particle balance:

$$\int_{E_g}^{\infty} \frac{E^2 dE}{\exp(\frac{E}{kT_1}) - 1} = \int_{E_g}^{\infty} \frac{E^2 dE}{\exp(\frac{E-\zeta_{oc}E_g}{kT_2}) - 1} \ ,$$

but no detailed balance. This is illustrated by the spectral curves on Fig. 7.1.

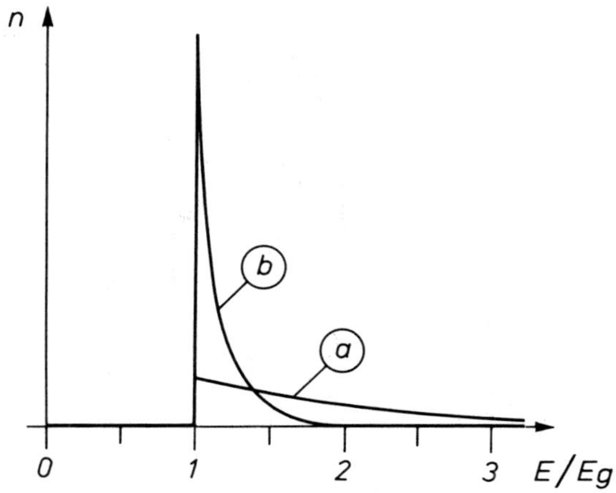

Fig. 7.1. Absorbed and emitted spectra of a solar energy converter under open-circuit condition: (a) incoming spectrum, which is also the outgoing spectrum in the photothermal case, (b) outgoing spectrum in the photovoltaic case.

We now calculate the entropy production rate. We have

$$\Sigma S(\zeta) = \frac{U(\zeta) - W(\zeta)}{T_2} - \frac{U(\zeta)}{T_1} ,$$

such that

$$\Sigma S(\zeta_{oc}) = (\frac{1}{T_2} - \frac{1}{T_1})\, U(\zeta_{oc})$$

and

$$\Sigma S(\zeta_{eq}) = -\, \frac{1}{T_2}\, W(\zeta_{eq}) .$$

Neither of the quantities $\Sigma S(\zeta_{oc})$ and $\Sigma S(\zeta_{eq})$ is zero. One can verify that both are positive. We see in Fig. 6.4(d) that for no value of ζ is the entropy creation rate $\Sigma S(\zeta)$ zero.

Thus, for $\zeta = \zeta_{eq}$ we have a non-zero entropy creation rate $\Sigma S(\zeta_{eq})$. This entropy creation is not caused by the existence of any energy transport, as $U(\zeta_{eq}) = 0$. We can associate it with the particle transport N. Equally, for $\zeta = \zeta_{oc}$ we have a non-zero entropy creation rate $\Sigma S(\zeta_{oc})$. This entropy creation is not caused by the existence of any particle transport, as $N(\zeta_{oc}) = 0$. We can associate it with the energy transport U. It is thus clear that a model is necessary where U and N are treated on an equal basis. Therefore, in Section 7.3 we will remodel endoreversibility, taking into account both energy and mass transports.

7.3 A unified model

From Section 7.2, it is clear that the 'driving' parameters η of photothermal conversion and ζ of photovoltaic conversion play similar roles. Therefore it is interesting to incorporate them in a single solar energy conversion model.

For this purpose, we consider again thermodynamic reservoirs, not defined merely by a temperature T, but by two parameters: their temperature T and their chemical potential μ, as in Section 6.2. The 'energy' and 'matter' fluxes U and N between reservoir (T_1, μ_1) and reservoir (T_3, μ_3) obey the following laws:

$$\begin{aligned} U &= g\,[\int_{E_g}^{\infty} \frac{E^3 dE}{\exp(\frac{E-\mu_1}{kT_1}) - 1} - \int_{E_g}^{\infty} \frac{E^3 dE}{\exp(\frac{E-\mu_3}{kT_3}) - 1}] \\ N &= g\,[\int_{E_g}^{\infty} \frac{E^2 dE}{\exp(\frac{E-\mu_1}{kT_1}) - 1} - \int_{E_g}^{\infty} \frac{E^2 dE}{\exp(\frac{E-\mu_3}{kT_3}) - 1}] . \end{aligned} \tag{7.1}$$

See Fig. 7.2(b).

These transport laws are complemented by the following three axioms:

Axiom 0: conservation of matter:

$$\Sigma N = 0 ,$$

where N are the particle flows and Σ denotes a summation over an arbitrary closed surface.

Axiom 1: conservation of energy:

$$\Sigma U + \Sigma W = 0 ,$$

where U are the energy flows, W the work flows and Σ denotes again a summation over an arbitrary closed surface.

Axiom 2: conservation of entropy:

$$\Sigma \frac{U - \mu N}{T} = 0 , \tag{7.2}$$

where Σ denotes a summation over a surface enclosing the reversible part only.

Axiom 2 expresses the reversibility of the engine between the reservoirs (T_3, μ_3) and (T_2, μ_2). It constitutes a generalized form of Axiom 2 of Section 3.2, i.e.

$$\Sigma \frac{Q}{T} = 0 .$$

See, for example, reference [1]. This means that an energy flow U is considered as build-up of two parts: a 'high-quality' part μN and a 'low-quality' part $Q = U - \mu N$. The former part is not accompanied by any entropy flow, the latter is accompanied by an entropy flow $S = Q/T$, such that a total energy flow U is accompanied by the entropy flow $(U - \mu N)/T$.

Taking Axioms 0 and 1 into account, eqn (7.2) becomes

$$\frac{U - W - \mu_2 N}{T_2} - \frac{U - \mu_3 N}{T_3} = 0 \ ,$$

which leads to

$$W = (1 - \frac{T_2}{T_3})U + (\frac{T_2}{T_3}\mu_3 - \mu_2)N \ .$$

If we postulate

$$W = \eta U + \zeta E_g N \ ,$$

we obtain a 'vectorial' efficiency (η, ζ), with

$$\eta = 1 - \frac{T_2}{T_3} \quad , \quad \zeta = \frac{T_2}{T_3}\frac{\mu_3}{E_g} - \frac{\mu_2}{E_g} \ . \tag{7.3}$$

If not only the engine part between reservoir 3 and reservoir 2 is reversible, but if the whole engine is reversible, i.e. when $T_3 = T_1$ and $\mu_3 = \mu_1$, then η and ζ take their reversible or Carnot values, given by

$$\eta_C = 1 - \frac{T_2}{T_1} \quad , \quad \zeta_C = \frac{T_2}{T_1}\frac{\mu_1}{E_g} - \frac{\mu_2}{E_g} \ . \tag{7.4}$$

In non-reversible conditions, the entropy creation rate is given by

$$\begin{aligned} \Sigma S(\eta, \zeta) &= \frac{(U - W) - \mu_2 N}{T_2} - \frac{U - \mu_1 N}{T_1} \\ &= \frac{U - \eta U - \zeta E_g N - \mu_2 N}{T_2} - \frac{U - \mu_1 N}{T_1} \ . \end{aligned}$$

Simple regrouping of terms leads to

$$\Sigma S(\eta, \zeta) = (1 - \frac{T_2}{T_1} - \eta)\frac{U}{T_2} + (\frac{T_2}{T_1}\mu_1 - \mu_2 - \zeta E_g)\frac{N}{T_2} \ .$$

With the help of eqn (7.3) this becomes

$$\Sigma S(\eta, \zeta) = (\frac{1}{T_3} - \frac{1}{T_1})U + (\frac{\mu_1}{T_1} - \frac{\mu_3}{T_3})N \ , \tag{7.5}$$

whereas with the help of eqn (7.4) it becomes

$$\Sigma S(\eta, \zeta) = (\eta_C - \eta)\frac{U}{T_2} + (\zeta_C - \zeta)\frac{E_g N}{T_2} \ .$$

The solar energy conversion is depicted in Fig. 7.2(c), derived from Fig. 7.2(b) by postulating $\mu_1 = 0$ and $\mu_2 = 0$. We now replace the parameters

T_3 and μ_3 by the dimensionless driving parameters $\eta = 1 - T_2/T_3$ and $\zeta = T_2\mu_3/T_3E_g = T_2qV/T_3E_g$. This changes eqns (7.1) into

$$U(\eta,\zeta) = g\,[\int_{E_g}^{\infty} \frac{E^3dE}{\exp(\frac{E}{kT_1})-1} - \int_{E_g}^{\infty} \frac{E^3dE}{\exp(\frac{E(1-\eta)-\zeta E_g}{kT_2})-1}\,]$$

$$N(\eta,\zeta) = g\,[\int_{E_g}^{\infty} \frac{E^2dE}{\exp(\frac{E}{kT_1})-1} - \int_{E_g}^{\infty} \frac{E^2dE}{\exp(\frac{E(1-\eta)-\zeta E_g}{kT_2})-1}\,].$$

The corresponding functions

$$W(\eta,\zeta) = \eta U + \zeta E_g N$$

$$\Sigma S(\eta,\zeta) = \frac{U-W}{T_2} - \frac{U}{T_1}$$

are plotted on Figs. 7.3 and 7.4, respectively. We see that W displays a maximum, located neither on the η-axis nor on the ζ-axis, and that ΣS displays a zero minimum on the η-axis.

In Figs. 7.3 and 7.4, the planes $\eta = 0$ and $\zeta = 0$ are displayed. The intersection of these planes with the curved surfaces W and ΣS generate the curves $W(\eta)$ and $\Sigma S(\eta)$ of Fig. 3.8 and the curves $W(\zeta)$ and $\Sigma S(\zeta)$ of Fig. 6.4.

Numerical calculation [7] of $w = W/\sigma T_1^4$ leads to the curve labelled pt/pv on Fig. 7.5(a), where it is compared with the corresponding curves, copied from Figs. 5.8 and 6.5.

In this chapter we have considered so far only the special case of full concentration: $C = 1/f$. Of course, models can be generalized for an arbitrary value C. Figure 7.5(b) gives the numerical results for $C = 1$.

From Fig. 7.5 we conclude that

- for unconcentrated sunlight ($C = 1$), maximum efficiency is $w =$ 67.5 per cent, compared to 53.6 per cent for purely photothermal conversion and 31.0 per cent for purely photovoltaic conversion.
- for fully concentrated sunlight ($C = 1/f$), maximum efficiency is w = 86.2 per cent, compared to 85.4 per cent for purely photothermal conversion and 40.8 per cent for purely photovoltaic conversion.

The reason why hybrid conversion leads to efficiencies higher than purely photothermal or purely photovoltaic conversion, is illustrated by Fig. 7.3: the pt/pv converter has two degrees of freedom, i.e. η and ζ (T_3 and V_3) and is therefore able to choose a better working point (η,ζ) than a pt system which is constrained to choose its working point on the line $\zeta = 0$, or a pv system which is constrained to remain on the line $\eta = 0$.

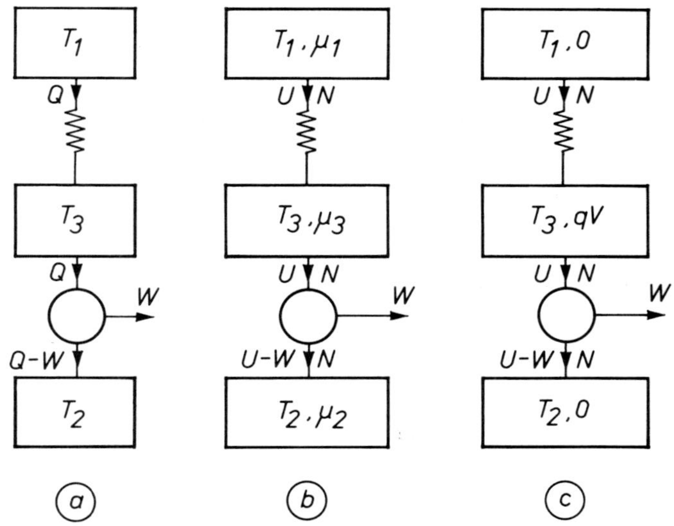

Fig. 7.2. Three thermodynamic configurations: (a) Müser configuration, (b) generalized configuration, (c) solar energy configuration (becoming Müser's configuration, if $V = 0$; becoming the solar cell configuration, if $T_3 = T_2$).

7.4 Alternative model ♡

In Section 7.3, we demonstrated that a generalized endoreversible engine enables us to describe both the photothermal and the photovoltaic converters by one and the same model. Here two parameters, namely η and ζ, are present. There is one single reversible operating point: $(\eta, \zeta) = (\eta_C, \zeta_C)$. The apparent asymmetry between photothermal and photovoltaic solar energy conversion is caused by the fact that in this particular case $\eta_C \neq 0$ and $\zeta_C = 0$, which in its turn is a consequence of the fact that both μ_1, the chemical potential of the Sun, and μ_2, the chemical potential of the planet, are zero.

If we now come back to the question posed in the title of Section 7.2, we can without doubt answer 'yes'. The original definition of endoreversible engines, introduced by Rubin, i.e. ' irreversible engines where all irreversibilities are restricted to the *coupling* of the engine to the external world ' can still be handled, provided we consider as 'coupling' not merely '*heat exchange*' but '*energy or matter exchange*'. Alternatively, one could consider '*heat or matter exchange*'. In the first case we consider as permit-

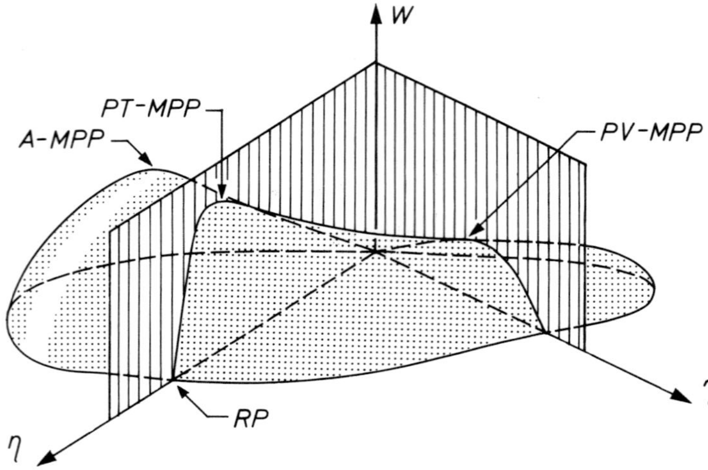

Fig. 7.3. Work production W as a function of η and ζ (A-MPP, absolute maximum-power point; PT-MPP, photothermal maximum-power point; PV-MPP, photovoltaic maximum-power point; RP, reversible point).

ted sources of irreversibilities the two flux types U and N; in the second case we consider the two flux types Q and N, as introduced by Callen [1]. However, whereas the fluxes U and N are divergenceless within a thermal conductor g, this is not always the case for fluxes Q. This makes models based on the two independent fluxes Q and N somewhat more complicated than models based on the two independent fluxes U and N. This phenomenon will be treated more extensively in Section 7.5, as well as in Chapter 9.

Instead of working with the two flows U and N:

$$\begin{aligned} U &= g\,[\int_{E_g}^{\infty} \frac{E^3 dE}{\exp(\frac{E}{kT_1}) - 1} - \int_{E_g}^{\infty} \frac{E^3 dE}{\exp(\frac{E-\mu_3}{kT_3}) - 1}] \\ N &= g\,[\int_{E_g}^{\infty} \frac{E^2 dE}{\exp(\frac{E}{kT_1}) - 1} - \int_{E_g}^{\infty} \frac{E^2 dE}{\exp(\frac{E-\mu_3}{kT_3}) - 1}]\,, \end{aligned}$$

we can work with the two flows Q and N, where Q is defined as

$$Q = U - \mu_3 N\ .$$

Hence we now have

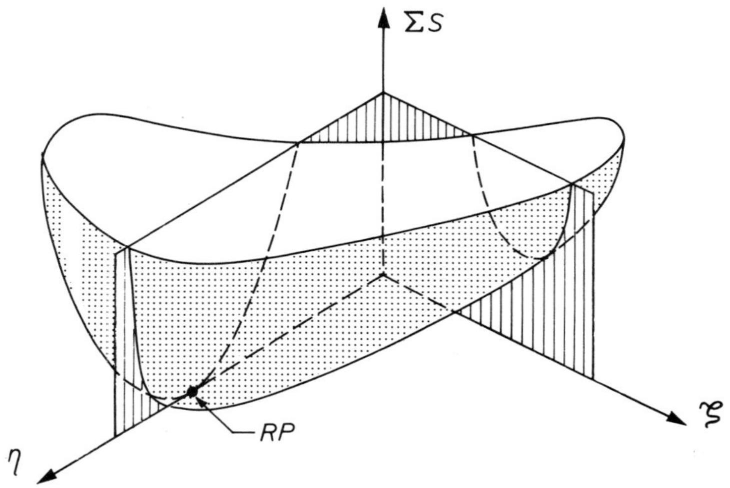

Fig. 7.4. Entropy creation rate ΣS as a function of η and ζ (RP, reversible point).

$$Q = g\left[\int_{E_g}^{\infty}\frac{(E-\mu_3)E^2dE}{\exp(\frac{E}{kT_1})-1}-\int_{E_g}^{\infty}\frac{(E-\mu_3)E^2dE}{\exp(\frac{E-\mu_3}{kT_3})-1}\right]$$

$$N = g\left[\int_{E_g}^{\infty}\frac{E^2dE}{\exp(\frac{E}{kT_1})-1}-\int_{E_g}^{\infty}\frac{E^2dE}{\exp(\frac{E-\mu_3}{kT_3})-1}\right].$$

Axiom 2 now yields

$$\frac{Q-W+(\mu_3-\mu_2)N}{T_2}-\frac{Q}{T_3}=0 .$$

This leads to

$$W=(1-\frac{T_2}{T_3})Q+(\mu_3-\mu_2)N .$$

If we now postulate

$$W=\eta Q+\lambda E_g N ,$$

then we obtain a new 'vectorial' efficiency (η,λ):

$$\eta=1-\frac{T_2}{T_3} \quad , \quad \lambda=\frac{\mu_3-\mu_2}{E_g} ,$$

which is an alternative for eqn (7.3). We remark that the present expression for λ is simpler than the corresponding expression for ζ in eqn (7.3).

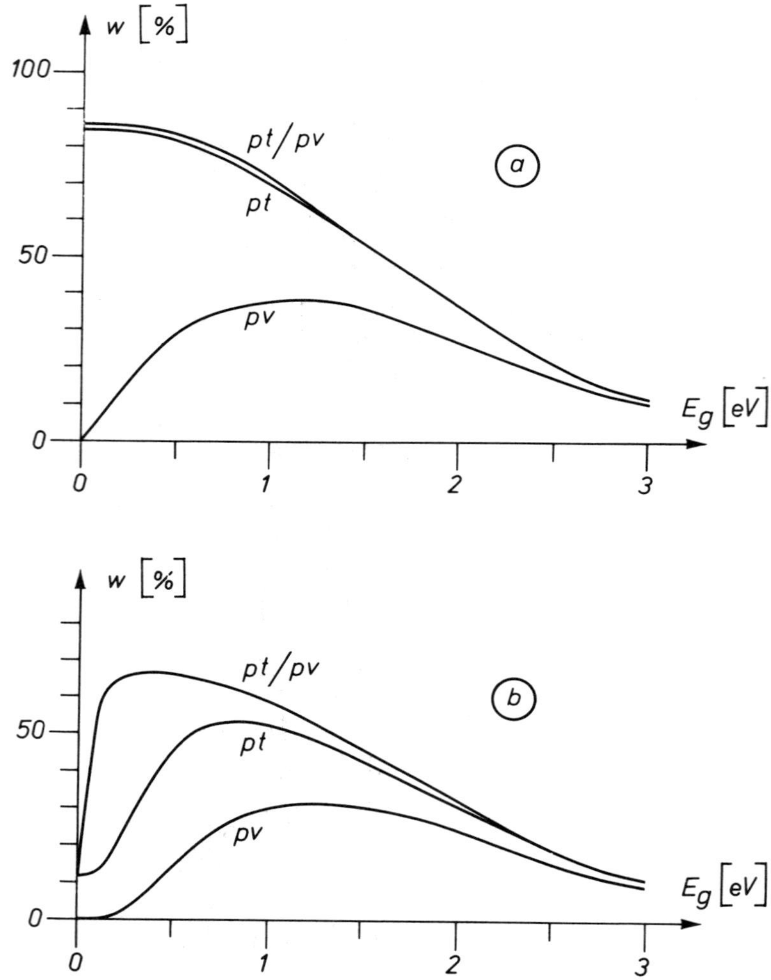

Fig. 7.5. Work produced by a hybrid photothermal/photovoltaic solar energy converter, compared to a purely photothermal and a purely photovoltaic converter, as a function of bandgap E_g: (a) with fully concentrated sunlight, (b) with non-concentrated sunlight.

Replacing the solar cell parameters T_3 and μ_3 by their dimensionless counterparts η and λ yields

$$\begin{aligned} Q(\eta,\lambda) &= g\,[\int_{E_g}^{\infty} \frac{(E-\lambda E_g)E^2 dE}{\exp(\frac{E}{kT_1})-1} - \int_{E_g}^{\infty} \frac{(E-\lambda E_g)E^2 dE}{\exp(\frac{(E-\lambda E_g)(1-\eta)}{kT_2})-1}] \\ N(\eta,\lambda) &= g\,[\int_{E_g}^{\infty} \frac{E^2 dE}{\exp(\frac{E}{kT_1})-1} - \int_{E_g}^{\infty} \frac{E^2 dE}{\exp(\frac{(E-\lambda E_g)(1-\eta)}{kT_2})-1}]\,, \end{aligned}$$

where we have taken into account that $\mu_2 = 0$.

Because

$$\begin{aligned} \eta &= 1 - \frac{T_2}{T_3}\,, \\ \zeta &= \frac{T_2}{T_3}\,\frac{\mu_3}{E_g}\,, \\ \lambda &= \frac{\mu_3}{E_g}\,, \end{aligned}$$

we can say that the new independent variables (η, λ) are related to the former independent parameters (η, ζ) by the (non-linear) transformation of coordinates:

$$\lambda = \frac{\zeta}{1-\eta}\,.$$

7.5 Onsager's principle ♡

Expression (7.5) for ΣS gives us the opportunity to tell something more about the functions, called function$_1$ and function$_2$ in eqn (6.2).

We first remark that in eqn (7.5) the parameters $1/T$ and μ/T play a fundamental role. Therefore we will consider them as the independent intensive variables instead of T and μ as before. If we define the two affinities x and y as follows:

$$\begin{aligned} x &= -\frac{1}{T} \\ y &= \frac{\mu}{T}\,, \end{aligned}$$

then we can write

$$\Sigma S = (x_1 - x_3)U + (y_1 - y_3)N\,, \tag{7.6}$$

where

$$\begin{aligned} U &= X(x_1, y_1) - X(x_3, y_3) \\ N &= Y(x_1, y_1) - Y(x_3, y_3)\,. \end{aligned}$$

Here, X and Y are functions, which we do not specify yet.

For any given couples (x_1, y_1) and (x_3, y_3), the above expression of ΣS must be positive or zero. If we apply this condition in particular to two closely spaced couples

$$\begin{aligned}(x_1, y_1) &= (x, y)\\ (x_3, y_3) &= (x + dx, y + dy)\ ,\end{aligned}$$

then eqn (7.6) becomes

$$\Sigma S = -dx(-\frac{\partial X}{\partial x}dx - \frac{\partial X}{\partial y}dy) - dy(-\frac{\partial Y}{\partial x}dx - \frac{\partial Y}{\partial y}dy)\ ,$$

yielding

$$\Sigma S = \frac{\partial X}{\partial x}(dx)^2 + (\frac{\partial X}{\partial y} + \frac{\partial Y}{\partial x})dxdy + \frac{\partial Y}{\partial y}(dy)^2\ .$$

If a quadratic expression

$$a\,(dx)^2 + b\ dxdy + c\,(dy)^2$$

has to be positive or zero, irrespective of the values of dx and dy, this can only be guaranteed if the expression is positive definite, i.e. if both $a \geq 0$ and $c \geq 0$ and moreover $b^2 - 4ac \leq 0$. This gives rise to

$$\frac{\partial X}{\partial x} \geq 0 \tag{7.7}$$

$$\frac{\partial Y}{\partial y} \geq 0 \tag{7.8}$$

$$(\frac{\partial X}{\partial y} + \frac{\partial Y}{\partial x})^2 - 4\,\frac{\partial X}{\partial x}\frac{\partial Y}{\partial y} \leq 0\ . \tag{7.9}$$

The first result simply means that X has to be an increasing function of x, i.e. an increasing function of T. This means that the energy flux U has to flow from high temperature to low temperature. Analogously the second result simply means that Y has to be an increasing function of y, i.e. an increasing function of μ. This means that the matter flux N has to flow from high chemical potential to low chemical potential. Transport laws for U and N which do not satisfy these two conditions obviously violate the second law of thermodynamics. The third result is concerned with cross-effects: it restricts the possible energy fluxes U caused by potential differences and matter fluxes N caused by temperature differences.

For our radiative energy and matter exchange (6.1) we have

$$\begin{aligned} X(x,y) &= g\int_{E_g}^{\infty} \frac{E^3 dE}{\exp(\frac{-Ex-y}{k})-1} \\ Y(x,y) &= g\int_{E_g}^{\infty} \frac{E^2 dE}{\exp(\frac{-Ex-y}{k})-1} \ . \end{aligned}$$

Therefore

$$\begin{aligned} \frac{\partial X}{\partial x} &= \frac{g}{k}\int_{E_g}^{\infty} \frac{\exp(\frac{-Ex-y}{k})E^4 dE}{[\exp(\frac{-Ex-y}{k})-1]^2} \\ \frac{\partial X}{\partial y} &= \frac{g}{k}\int_{E_g}^{\infty} \frac{\exp(\frac{-Ex-y}{k})E^3 dE}{[\exp(\frac{-Ex-y}{k})-1]^2} \\ \frac{\partial Y}{\partial x} &= \frac{g}{k}\int_{E_g}^{\infty} \frac{\exp(\frac{-Ex-y}{k})E^3 dE}{[\exp(\frac{-Ex-y}{k})-1]^2} \\ \frac{\partial Y}{\partial y} &= \frac{g}{k}\int_{E_g}^{\infty} \frac{\exp(\frac{-Ex-y}{k})E^2 dE}{[\exp(\frac{-Ex-y}{k})-1]^2} \ . \end{aligned}$$

The reader will easily verify that the above expressions fulfil the two conditions (7.7) and (7.8) with the inequality sign and fulfil the condition (7.9) with the equality sign.

One of the fundamental laws of non-equilibrium thermodynamics, not far from equilibrium, is called Onsager's relation. It expresses one more relationship between the partial derivatives:

$$\frac{\partial X}{\partial y} = \frac{\partial Y}{\partial x} \ . \tag{7.10}$$

Note that it restricts the cross-effects by an equality condition, not by merely a $\leq$ relation. The reader will easily verify that also the present condition is fulfilled by the radiative rate equations.

7.6 An example

Examples of hybrid photothermal/photovoltaic systems are not easy to find. We describe here a building at the Brown University Campus at Providence (Rhode Island, U.S.A.). Its roof is covered with 34 m^2 panels, containing 2860 circular silicon solar cells, 102 mm in diameter. See Fig. 7.6.

Whereas the solar cells provide electricity, an air conduct system extracts heat from the panels. The intermediate temperature T_3 equals 55

Fig. 7.6. Brown University pt/pv house.

to 75 °C. However, the hot air is not lead to a Carnot thermal engine for producing work, but instead is used for home heating [3] [4].

Although some pt/pv systems including a Carnot engine (or other heat engine, such as Rankine engine) have been designed, no demonstration project has (yet) been built.

Bibliography

1. H. Callen (1960). *Thermodynamics.* Wiley, New York, pp. 45 - 46 and 293 - 296.

2. J. Williams (1977). *Solar energy: technology and applications.* Ann Arbor Publishers, Ann Arbor, pp. 98 - 102.

3. J. Loferski, C. Case, G. Doodlesack, B. Roessler, R. Dobbins, T. Russell, J. Beall and J. Krikorian (1982). Design and construction of a hybrid 3 kWp photovoltaic - thermal solar energy system for residential - commercial building. *Proceedings of the 16 th I.E.E.E. Photo-*

voltaic Specialists Conference, San Diego, 27 - 30 September 1982, pp. 188 - 193.

4. J. Loferski, J. Ahmad and A. Pandey (1988). Performance of photovoltaic cells incorporated into unique hybrid photovoltaic/thermal panels of a 2.8 kW residential solar energy conversion system. *Proceedings of the 1988 annual meeting of the American Solar Energy Society*, Cambridge, 20 - 24 June 1988, pp. 427 - 432.

5. R. Haase (1990). *Thermodynamics of irreversible processes.* Dover, New-York, pp. 88 - 92, 110 - 111 and 167 - 172.

6. A. De Vos (1991). Is a solar cell an endoreversible engine? *Solar Cells*, volume 31, pp. 181 - 196.

7. A. De Vos and J. Landries (1992). Endoreversible thermodynamics of the hybrid photothermal/photovoltaic converter. In press.

8
Multicolour conversion

8.1 Introduction

In Chapters 5 (pt conversion), 6 (pv conversion) and 7 (pt/pv conversion) we applied a single solid-state material. The latter was characterized by one particular material constant: the bandgap E_g.

Human ingenuity, however, goes further: one can conceive solar energy converters, built up from two (or more) materials, each with its own bandgap. Such systems are sometimes referred to as tandem systems.

We will see, in this chapter, that in such systems, each material is responsible for the conversion of a particular energy band of the incoming solar irradiance. In other words, each subsystem absorbs and converts part of the solar spectrum. Because any given slice of a light spectrum is characterized by a particular colour, we can say that each subsystem is responsible for the energy conversion of a particular colour. Therefore multibandgap systems are also called multicolour systems. Analogously, we call the single gap systems of Chapters 5, 6 and 7 one-colour systems.

In this chapter, we will consecutively describe multicolour pt converters and multicolour pv converters. We will not describe multicolour pt/pv, i.e. hybrid converters, as the reader can easily construct such a generalization.

A challenging problem is generated if one lets n, i.e. the number of applied materials and thus the number of different bandgaps, increase without limit: $n \to +\infty$. Such systems, which we will call omnicolour systems, will also be considered.

8.2 Photothermal multicolour converters

We have seen in Chapter 5 how (for $Cf < 1$) the Müser engine can be improved by replacing the black-body solar absorber by a selectively black absorber, such that

$$\epsilon(E) = \begin{cases} 0 & \text{for } E < E_g \\ 1 & \text{for } E > E_g \,. \end{cases}$$

In this chapter we go a step further by applying n different materials, each with its own bandgap E_{gi}: using selective materials obeying

$$\epsilon_i(E) = \begin{cases} 0 & \text{for } E < E_{gi} \\ 1 & \text{for } E > E_{gi} \end{cases} ,$$

with

$$1 \leq i \leq n \; .$$

We stack those absorbers one above the other, with increasing bandgap. The largest bandgap thus faces the Sun. That one, we will label as 1, and so on, until the bottom absorber, i.e. the one with smallest bandgap, as n: see Fig. 8.1. We thus have

$$E_{g1} > E_{g2} > \ldots > E_{gn} \; .$$

Sunlight with energy E larger than E_{g1} is absorbed by the uppermost absorption layer. Photons with smaller energy travel through the first absorber and enter the second layer. There, all the photons are absorbed with energy larger than E_{g2} (and smaller than E_{g1}, as no photons with $E > E_{g1}$ are incident on the second absorber). This second layer is transparent for all light with $E < E_{g2}$. Such photons reach absorber 3, etc. Thus the ith unit absorbs all solar photons with energy in the range $(E_{gi}, E_{g(i-1)})$, except for $i = 1$, as the first layer absorbs all photons with energy in the range $(E_{g1}, +\infty)$. Note finally that all photons with energy smaller than the smallest band gap E_{gn} are reflected by the metal-coated bottom side of the bottom converter.

As in Chapter 7, we will restrict formulas in this chapter to the case of fully concentrated sunlight, i.e. $Cf = 1$.

If each absorber has an own temperature T_i, then the ith absorber emits an energy spectrum

$$\begin{cases} 0 & \text{for } E < E_{gi} \\ 2g \, \dfrac{E^3}{\exp(E/kT_i) - 1} & \text{for } E > E_{gi} \; . \end{cases}$$

Three spectra are incident on the absorber:

- Part of the solar spectrum

$$\begin{cases} g \, \dfrac{E^3}{\exp(E/kT_s) - 1} & \text{for } E < E_{g(i-1)} \\ 0 & \text{for } E > E_{g(i-1)} \; ; \end{cases}$$

  only a part of this part is actually absorbed:

$$\begin{cases} 0 & \text{for } E < E_{gi} \\ g \, \dfrac{E^3}{\exp(E/kT_s) - 1} & \text{for } E_{gi} < E < E_{g(i-1)} \\ 0 & \text{for } E > E_{g(i-1)} \; . \end{cases}$$

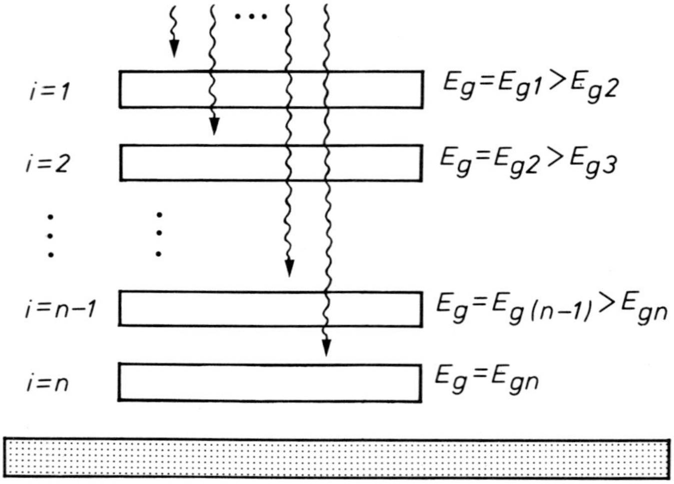

Fig. 8.1. Tandem stack.

- The spectrum emitted by the layer above:

$$\begin{cases} 0 & \text{for } E < E_{g(i-1)} \\ g\,\dfrac{E^3}{\exp(E/kT_{i-1})-1} & \text{for } E > E_{g(i-1)} \;; \end{cases}$$

  this spectrum is completely absorbed.

- The spectrum emitted by the layer beneath:

$$\begin{cases} 0 & \text{for } E < E_{g(i+1)} \\ g\,\dfrac{E^3}{\exp(E/kT_{i+1})-1} & \text{for } E > E_{g(i+1)} \;; \end{cases}$$

  only a part of this spectrum is actually absorbed:

$$\begin{cases} 0 & \text{for } E < E_{gi} \\ g\,\dfrac{E^3}{\exp(E/kT_{i+1})-1} & \text{for } E > E_{gi} \;. \end{cases}$$

As a result, the net heat flux Q_i absorbed by layer i is given by

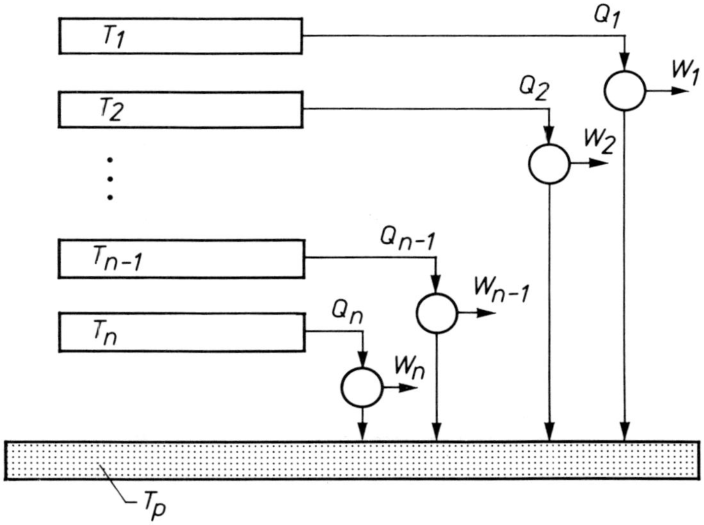

Fig. 8.2. Photothermal tandem converter.

$$Q_i = g \Big[\int_{E_{gi}}^{E_{g(i-1)}} \frac{E^3 dE}{\exp(\frac{E}{kT_s}) - 1} + \int_{E_{g(i-1)}}^{\infty} \frac{E^3 dE}{\exp(\frac{E}{kT_{i-1}}) - 1} + \int_{E_{gi}}^{\infty} \frac{E^3 dE}{\exp(\frac{E}{kT_{i+1}}) - 1} - 2\int_{E_{gi}}^{\infty} \frac{E^3 dE}{\exp(\frac{E}{kT_i}) - 1} \Big] . \quad (8.1)$$

Note that this expression is similar to eqn (5.15) with two additional terms (i.e. the spectra incident from the two neighbouring layers) and an additional factor 2 in the last term.

Now each absorber is connected to the hot side of its own Carnot engine. Finally the cold sides of all Carnot engines are connected to the same cold reservoir: the surroundings at temperature T_p. See Fig. 8.2.

Each Carnot engine produces a power W_i. The total work delivered per unit time is

$$W = \Sigma_i W_i = \Sigma_i (1 - \frac{T_p}{T_i}) Q_i .$$

We will not go into the details of the calculation. Suffice it here to give the results [6]: Fig. 8.3 shows the maximum deliverable work as a function of

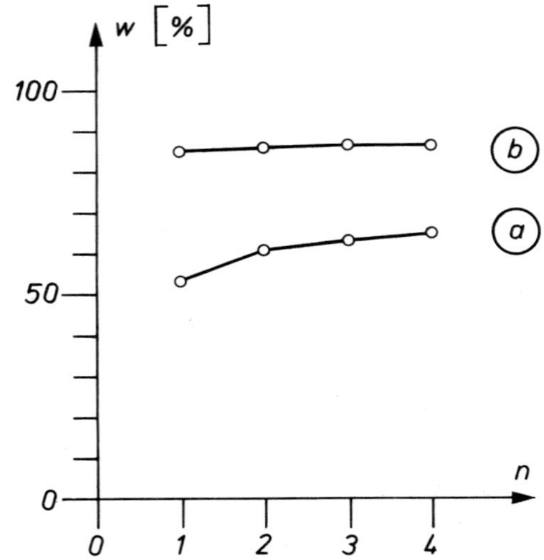

Fig. 8.3. Multicolour photothermal solar energy conversion efficiency: (a) with non-concentrated sunlight, (b) with fully concentrated sunlight.

n, the number of layers in the system. We see on curve (a) how w increases with increasing n for natural sunlight ($C = 1$). For fully concentrated sunlight ($Cf = 1$), the curve $w(n)$ has such a small slope, that the increase is hardly noticeable: $w = 85.4$ per cent for $n = 1$, $w = 86.1$ per cent for $n = 2$, $w = 86.3$ per cent for $n = 3$, etc.

8.3 Photovoltaic multicolour converters

Like photothermal converters, solar cells with different bandgaps can be stacked: see Fig. 8.4. Each diode has its own maximum-power voltage V_i, just as each layer of the photothermal stack had its own maximum-power temperature T_i. The total produced power is of course given by $W = \Sigma_i W_i = \Sigma_i V_i I_i$.

Figure 8.5 shows how the efficiency of the multicolour system increases with increasing complexity n. We see how the multicolour photovoltaic converter efficiency grows faster, with increasing n, than the multicolour photothermal converter efficiency.

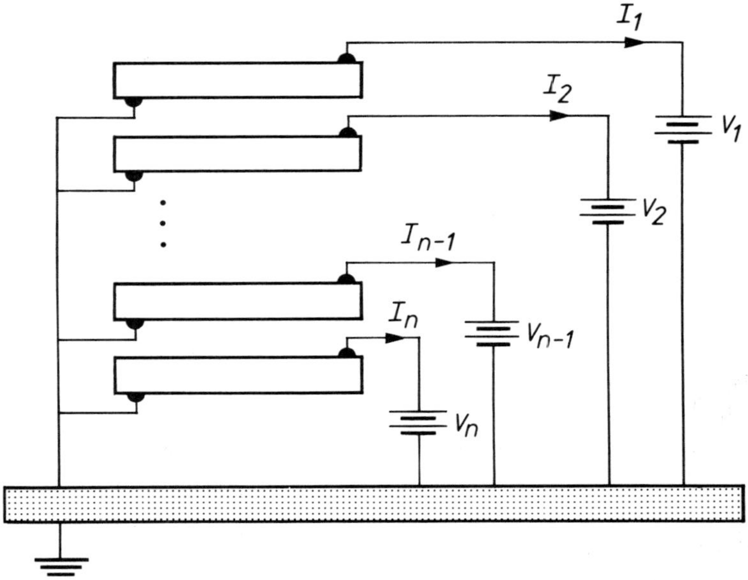

Fig. 8.4. Photovoltaic tandem converter.

8.4 Omnicolour converters

If we let n, the number of layers stacked in the photothermal converter, grow without limit, we find that the optimal bandgaps E_{gi} come closer and closer to one another. For $n \to +\infty$ we obtain an infinite stack, where E_g varies smoothly from $+\infty$ at the top to 0 at the bottom. The bandgap increment $E_{g(i-1)} - E_{gi}$ between two adjacent semiconductors becomes an infinitely small increment dE_g. See Fig. 8.6.

The power absorbed by a single semiconductor becomes infinitesimal:

$$dQ = g \left[\frac{E^3 dE}{\exp(\frac{E}{kT_s}) - 1} - \frac{E^3 dE}{\exp(\frac{E}{kT(E)}) - 1} \right] ,$$

where the last three terms of expression (8.1) have merged into one single term. Note that we have replaced E_g and dE_g by E and dE respectively. This is justified, as the energy E of the absorbed and emitted photons satisfies

$$E_g < E < E_g + dE_g \ ,$$

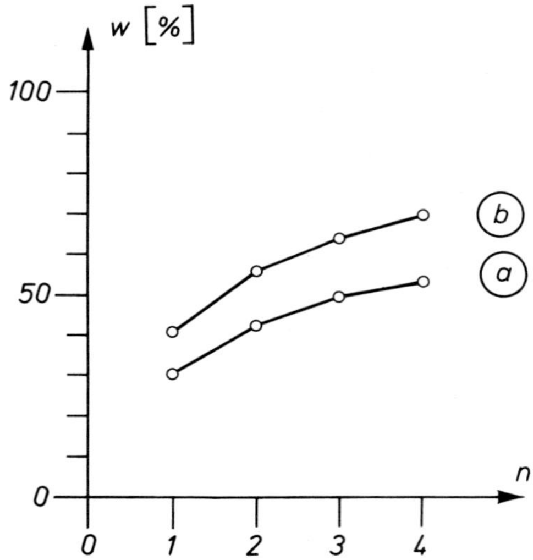

Fig. 8.5. Multicolour photovoltaic solar energy conversion efficiency: (a) with non-concentrated sunlight, (b) with fully concentrated sunlight.

so that, in the limit, there is no distinction any more between bandgap energy E_g and photon energy E. In the limit, the absorbers behave as two-level systems instead of as two-band systems.

As each layer is connected with a Carnot engine with efficiency

$$\eta(E) = 1 - \frac{T_p}{T(E)} \ ,$$

the total power is eventually given by

$$W = g \int_0^{+\infty} \eta(E) \left[\frac{1}{\exp(\frac{E}{kT_s}) - 1} - \frac{1}{\exp(\frac{E(1-\eta(E))}{kT_p}) - 1} \right] E^3 dE \ .$$

Note that the summation Σ_i of the partial powers W_i has become a smooth integral over the photon energy E. The function $\eta(E)$ in the above integral has to be chosen such that the integrand is maximized.

Division by the incoming power σT_s^4 finally yields the solar energy conversion efficiency

$$w = \frac{15}{\pi^4} \left(\frac{T_p}{T_s}\right)^4 \int_0^{+\infty} \eta \left[\frac{1}{\exp(\frac{T_p}{T_s}x) - 1} - \frac{1}{\exp(x(1-\eta)) - 1} \right] x^3 dx \ , \quad (8.2)$$

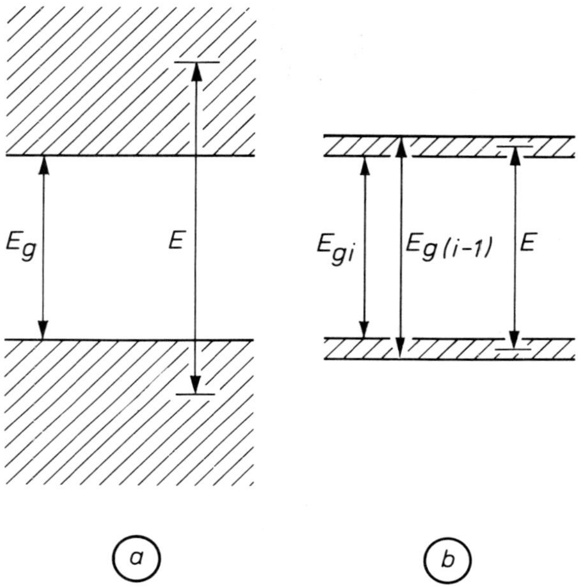

Fig. 8.6. Electron levels, able to exchange photons with the Sun: (a) one-colour system, (b) omnicolour system.

where η is the solution of the transcendental equation

$$\frac{d}{d\eta}\,\eta\;[\;\frac{1}{\exp(\frac{T_p}{T_s}x)-1}-\frac{1}{\exp(x(1-\eta))-1}\;]=0\;,$$

i.e.

$$\frac{(1+\eta x)\exp(x(1-\eta))-1}{[\exp(x(1-\eta))-1]^2}=\frac{1}{\exp(\frac{T_p}{T_s}x)-1}\;. \tag{8.3}$$

For sufficiently small values of T_p/T_s, we find, after laborious calculations [5], the following asymptotic expansion:

$$w=1+B_1\frac{T_p}{T_s}+C_1\frac{T_p}{T_s}\log(\frac{T_p}{T_s})+B_2(\frac{T_p}{T_s})^2+C_2(\frac{T_p}{T_s})^2\log(\frac{T_p}{T_s})+\dots\;, \tag{8.4}$$

where the first four coefficients are given by

$$
\begin{aligned}
B_1 &= -\frac{11}{12} - \frac{15}{\pi^4}\,(5-2\gamma)\,\zeta(3) - \frac{30}{\pi^4}\,\zeta'(3)\\
C_1 &= \frac{30}{\pi^4}\,\zeta(3)\\
B_2 &= \frac{5}{2\pi^2}\,(2-\gamma) - \frac{75}{4\pi^4} + \frac{15}{\pi^4}\,\zeta(3) + \frac{15}{\pi^4}\,\zeta'(2)\\
C_2 &= \frac{15}{2\pi^4} - \frac{5}{2\pi^2}\ .
\end{aligned}
$$

Apart from the familiar Riemann zeta function ζ, we also find here ζ', the first derivative of this zeta function. We also encounter γ, a mathematical constant, called the constant of Euler–Mascheroni which equals 0.57722. Because the expressions for B_1, etc. are so horribly complicated, we give here the numerical values:

$$
\begin{aligned}
B_1 &= - \ 1.56748\\
C_1 &= \ 0.37021\\
B_2 &= \ 0.20864\\
C_2 &= - \ 0.17631\ .
\end{aligned}
$$

For

$$\frac{T_p}{T_s} = \frac{288 \text{ K}}{5762 \text{ K}} = 0.0500\ ,$$

numerical calculation of eqns (8.2)-(8.3) yields $w = 0.86821$, whereas series (8.4), truncated after the C_2-term, yields $w = 0.86802$.

Figure 8.7(c) shows results of calculations for arbitrary T_p/T_s. They are compared to the curves 5.9(a) and 6.6(a) of one-colour systems.

When we apply the limit operation $n \to +\infty$ to the photovoltaic multicolour stack, we end up with an integral similar to eqns (8.2)-(8.3). After some inspection, the new integral can be transformed and turns out to be identical to eqns (8.2)-(8.3). Thus, in the limit, photothermal and photovoltaic conversion have the same efficiency. This is in agreement with Figs. 8.3 and 8.5, where we see that, for increasing n, the photovoltaic efficiency catches up with the photothermal efficiency.

This mathematical result is not a surprise, when we look for a physical explanation. Indeed, in the limit, each stack layer works as a two-level system. In a two-level system, there is no distinction between the photothermal spectrum

$$\frac{E^3\,dE}{\exp\left(\frac{E}{kT(E)}\right) - 1}$$

and the photovoltaic spectrum

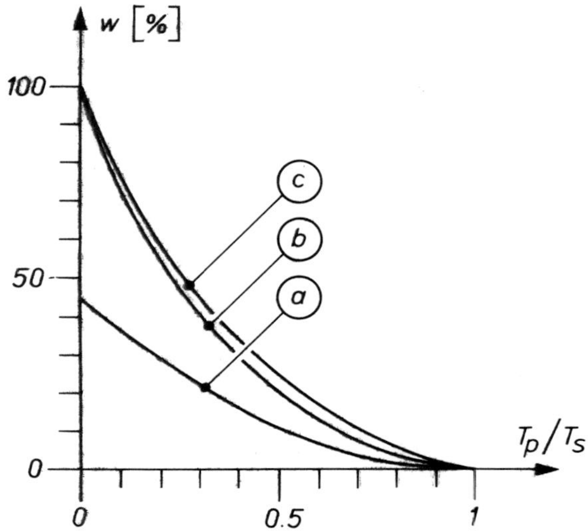

Fig. 8.7. One-colour and omnicolour conversion efficiencies ($C = C_{max} = 1/f$): (a) one-colour photovoltaic conversion, (b) one-colour photothermal conversion, (c) omnicolour conversion.

$$\frac{E^3 dE}{\exp(\frac{E-qV(E)}{kT_p}) - 1} \, .$$

It suffices to choose the functions $T(E)$ and $V(E)$ such that

$$\frac{E}{kT(E)} = \frac{E - qV(E)}{kT_p} \, .$$

8.5 The Bose engine ♡

In the omnicolour photothermal converter, each partial convertor is an endoreversible engine with a finite conductance for the input heat Q_1 :

$$Q_1 = gE^3 dE \left[\frac{1}{\exp(\frac{E}{kT_1}) - 1} - \frac{1}{\exp(\frac{E}{kT_3}) - 1} \right] .$$

This constitutes a special case of a broader class of endoreversible engines, which obey the heat transport equations

$$Q_1 = g_1 \left[\frac{1}{\exp(\frac{E}{kT_1}) - 1} - \frac{1}{\exp(\frac{E}{kT_3}) - 1} \right] \tag{8.5}$$

$$Q_2 = g_2 \left[\frac{1}{\exp(\frac{E}{kT_4}) - 1} - \frac{1}{\exp(\frac{E}{kT_2}) - 1} \right] . \tag{8.6}$$

We will call such engines Bose engines.

The above non-linear transport equations have to be compared with the linear transport equations of the Curzon–Ahlborn engines:

$$\begin{aligned} Q_1 &= g_1(T_1 - T_3) \\ Q_2 &= g_2(T_4 - T_2) \end{aligned}$$

and with the non-linear transport equations of the Stefan–Boltzmann engines:

$$\begin{aligned} Q_1 &= g_1(T_1^4 - T_3^4) \\ Q_2 &= g_2(T_4^4 - T_2^4) . \end{aligned}$$

Equations (8.5)-(8.6), together with the equations

$$\begin{aligned} \frac{Q_1}{T_3} &= \frac{Q_2}{T_4} \\ \eta &= 1 - \frac{T_4}{T_3} \end{aligned}$$

determine completely the current-efficiency characteristic $Q_1(\eta)$. Unfortunately, solving analytically the set of the two equations

$$\frac{g_1}{T_3} \left[\frac{1}{\exp(\frac{E}{kT_1}) - 1} - \frac{1}{\exp(\frac{E}{kT_3}) - 1} \right] = \frac{g_2}{T_4} \left[\frac{1}{\exp(\frac{E}{kT_4}) - 1} - \frac{1}{\exp(\frac{E}{kT_2}) - 1} \right]$$

$$\eta = 1 - \frac{T_4}{T_3}$$

for T_3 and T_4 is not possible. It is therefore no surprise that we cannot construct an explicit form for $Q_1(\eta)$. The best we can do, is to find an implicit relation between Q_1 and η:

$$\log\left[1 + \frac{1}{\frac{1}{\exp(\frac{E}{kT_1})-1} - \frac{Q_1}{g_1}} \right]$$

$$= (1 - \eta) \log\left[1 + \frac{1}{\frac{1}{\exp(\frac{E}{kT_2})-1} + (1 - \eta)\frac{Q_1}{g_2}} \right] .$$

Table 8.1. Multicolour solar energy conversion efficiency for natural sunlight ($C = 1$), moderately concentrated sunlight ($C = 100$) and fully concentrated sunlight ($C = 1/f$).

C	n	Photothermal	Photovoltaic
1	1	53.6	31.0
	2	60.9	42.9
	3	63.3	49.3
	...		
	∞	68.2	68.2
100	1	67.0	35.2
	2	71.7	48.4
	3	73.2	55.6
	...		
	∞	76.2	76.2
46 300	1	85.4	40.8
	2	86.1	55.7
	3	86.3	63.9
	...		
	∞	86.8	86.8

Therefore no general conclusions can be drawn about the form of $Q_1(\eta)$ or about the maximum power, which has to be deduced from

$$\frac{d}{d\eta}\,\eta Q_1(\eta) = 0\ .$$

That is why we will not study the general Bose engine any further.

8.6 Conclusion

Table 8.1 recalls the most important results of this chapter. We can conclude that multicolour systems attain higher efficiencies than one-colour systems. The gain is, however, more important for photovoltaic conversion than for photothermal conversion.

For $n \to +\infty$, any distinction between pv and pt (as well as hybrid) converters disappears. Combining $n = \infty$ with the maximal realizable light concentration, i.e. $C = 1/f$, yields the upper limit of any solar energy conversion efficiency theoretically possible. For the case of the Earth's environment, this limit is given by w = 86.8 per cent.

8.7 An example

An experimental multicolour device is shown in Fig. 8.8. It is a 20 mm^2 tandem solar cell fabricated by the Varian Research Center at Palo Alto (California, U.S.A.). It consists of three diodes made of the semiconductors $Al_xGa_{1-x}As$ (E_{g1} = 1.93 eV), GaAs (E_{g2} = 1.42 eV) and $In_xGa_{1-x}As$ (E_{g3} = 1.02 eV). It is designed to function on board space-crafts, together with a concentrator C = 100 - 1000. The three cells have been tested both on Earth and in Space [8] [9]. The latter experiment was performed on the satellite 'Lips 3', launched into a circular orbit, 1100 km above the Earth, in spring 1987.

The partial performances, at C = 100, amount to:

- w_1 = 15.4 per cent,
- w_2 = 10.8 per cent, and
- w_3 = 3.0 per cent,

such that the overall performance is w = 29.2 per cent. If we compare this number with the value w = 55.6 per cent of Table 8.1, we have to conclude that the theoretical bound is far from being realized. The cause is twofold:

- Firstly, each subcell suffers from the same drawbacks as the single gap solar cell. Those drawbacks were enumerated in Section 6.4 and can be summarized as follows:

  * not every photon with energy E in the range $(E_{gi}, E_{g(i-1)})$, that is incident on the ith subcell, gives rise to an electron jump from the valence band to the conduction band;

  * not every electron falling from the conduction band to the valence band gives rise to a photon.

- Secondly, the stacking of the various subcells gives rise to additional losses: cells 1 to $(i-1)$ are not completely transparent to photons with energies in the range $(E_{gi}, E_{g(i-1)})$, such that not all photons from this spectral slice reach the ith subcell.

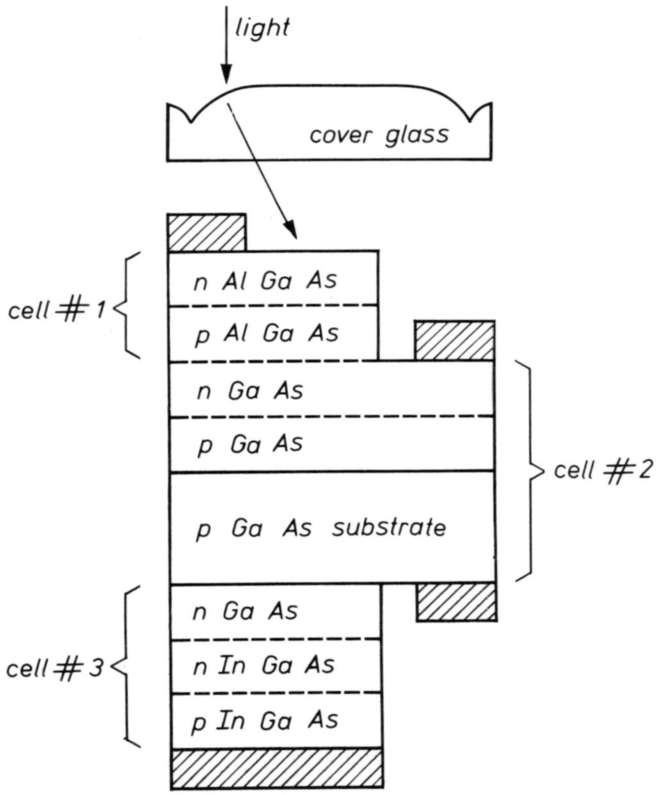

Fig. 8.8. Multicolour solar cell. Note: shaded areas are ohmic contacts.

Bibliography

1. A. Haught (1980). *Physics considerations of solar energy conversion.* United Technologies Research Center, internal report no. UTRC 80-9.

2. A. Haught (1980). Physics considerations of solar energy conversion. *Proceedings of the 3 rd International Conference on Photochemical Conversion and Storage of Solar Energy*, Boulder, 3 - 8 August 1980, pp. 443 - 445.

3. A. De Vos (1980). Detailed balance limit of the efficiency of tandem solar cells. *Journal of Physics D: Applied Physics*, volume 13, pp. 839 - 846.

4. H. Pauwels and A. De Vos (1980). Determination of the maximum efficiency solar cell structure. *Solid-State Electronics*, volume 24, pp. 835 - 843.

5. A. De Vos, C. Grosjean and H. Pauwels (1982). On the formula for the upper limit for photovoltaic solar energy conversion efficiency. *Journal of Physics D: Applied Physics*, volume 15, pp. 2003 - 2015.

6. A. De Vos and D. Vyncke (1983). Solar energy conversion: photovoltaic *versus* photothermal conversion. *Proceedings of the 5 th E.C. Photovoltaic Solar Energy Conference*, $A\theta\iota\nu\alpha\iota$, 17 - 21 October 1983, pp. 186 - 190.

7. P. Baruch (1985). A two-level system as a model for a photovoltaic cell. *Journal of Applied Physics*, volume 57, pp. 1347 - 1355.

8. L. Partain, B. Chung, G. Virshup, J. Schultz, H. Mac Millan, M. Ristow, M. Kuryla, K. Bertness, M. Klausmeier-Brown, M. Wanlass and T. Coutts (1990). Progress towards 30 % efficient, monolithic, three-junction, two-terminal concentrator solar cell for space applications. *Proceedings of the 21 th I.E.E.E. Photovoltaic Specialists Conference*, Kissimee, 21 - 25 May 1990, pp. 184 - 189.

9. M. Klausmeier-Brown, B. Chung, M. Ristow, N. Kaminar, J. Werthen and G. Virshup (1990). Multi-junction component cells on Lisp III. *Proceedings of the 21 th I.E.E.E. Photovoltaic Specialists Conference*, Kissimee, 21 - 25 May 1990, pp. 1153 - 1158.

9
Chemical reactions

9.1 Introduction

We demonstrated in Chapter 7 that a solar cell cannot be modelled as an endoreversible engine, unless heat reservoirs (i.e. reservoirs supplying heat Q, while keeping their temperature T constant) are generalized into heat-and-matter reservoirs (i.e. reservoirs supplying both heat Q and matter N, while keeping constant both their temperature T and their chemical potential μ), and heat conductors are generalized into heat-and-matter conductors. See Fig. 9.1(b).

To illustrate the usefulness of this extended definition, this chapter will discuss a model, in which pure matter reservoirs are in contact with a reversible energy converter through matter flux resistances. The model is relevant to chemical reactions. The matter reservoir at high chemical potential μ_1 is the reactant reservoir, the convertor is the reaction chamber and the matter reservoir at low chemical potential μ_2 is the reservoir collecting the reaction product. See Fig. 9.1(c).

This chapter is not concerned with solar energy directly. The concepts developed will, however, be useful in Chapter 10, where we will consider the conversion of solar energy into chemical energy (by means of photosynthesis).

9.2 Reversible and irreversible chemical engines

Figure 9.2(a) shows the basic features of the reversible chemical reactor:

- two particle reservoirs: one of the chemical compound A_1 with 'high' chemical potential μ_1 and one of the chemical compound A_2 with 'low' chemical potential μ_2;
- a reversible reactor between these two reservoirs, where the chemical reaction

$$A_1 \rightleftharpoons A_2$$

takes place.

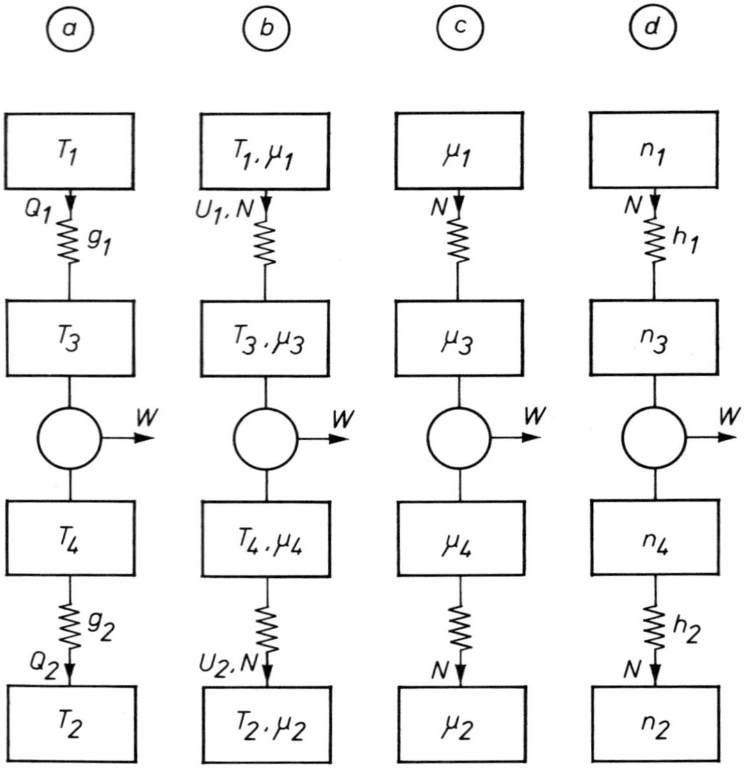

Fig. 9.1. Four thermodynamic engines: (a) with heat currents, (b) with energy and particle currents, (c) with particle currents, (d) with particle currents.

The adjectives 'high' and 'low' refer to the 'normal' operation, where μ_1 is higher than μ_2. However, we will not exclude the case $\mu_1 < \mu_2$.

In order to determine the properties of the structure, we have to implement the following two axioms only:

Axiom 0: conservation of matter:

$$\Sigma N = 0 \ ,$$

where N are the matter flows and Σ denotes a summation over a closed surface around the engine.

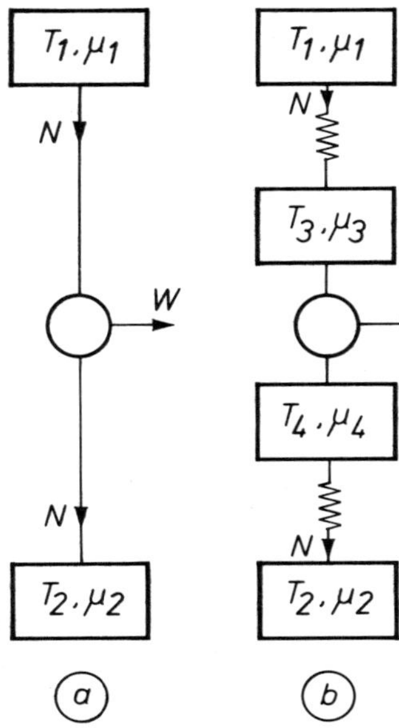

Fig. 9.2. Two thermodynamic configurations: (a) reversible reactor, (b) endoreversible reactor.

Axiom 1: conservation of energy:

$$\Sigma\mu N + \Sigma W = 0 \ ,$$

where W are the work flows.

If we introduce the following definition of the conversion efficiency ζ:

$$W = \zeta N \ ,$$

then Axiom 1 leads to

$$\zeta = \mu_1 - \mu_2 \ ,$$

which is the equivalent of Carnot's formula $\eta = (T_1 - T_2)/T_1$.

The reader should note that here the parameter ζ has the dimension of energy. It is similar to the ζ of Chapters 6 and 7, which, however, was dimensionless. In fact, the ζ of this Chapter plays the role of the ζE_g of the photovoltaic chapters.

We have to distinguish two modes of operation of the reversible engine:

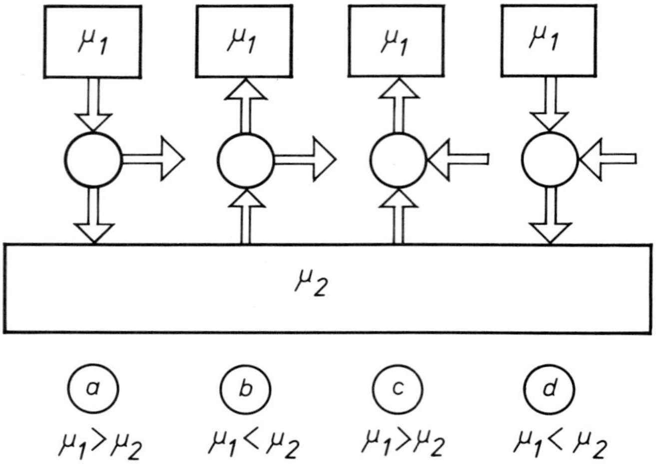

Fig. 9.3. Four thermodynamic modes: (a) chemical engine, (b) chemical engine, (c) A_1 pump, (d) A_2 pump. Note: all arrows indicate positive flows.

- If $W > 0$, then the engine delivers a positive work flow; it is said to function as a chemical engine.
- If $W < 0$, then the engine delivers a negative work flow. In other words, we have to supply a positive work flow $-W$ to the engine. We now distinguish two subcases:

    * If the 'surroundings' at potential μ_2 is the 'lower' reservoir (i.e. $\mu_1 > \mu_2$), then the engine is said to function as an A_1 pump: $\zeta > 0$ and thus $N < 0$. The work we supply is used to fill up reservoir 1, by sustaining the reaction $A_2 \rightarrow A_1$, instead of the spontaneous reaction $A_1 \rightarrow A_2$.
    * If the 'surroundings' at potential μ_2 is the 'higher' reservoir (i.e. $\mu_1 < \mu_2$), then the engine is said to function as an A_2 pump: $\zeta < 0$ and thus $N > 0$. The work we supply is used to empty reservoir 1, by sustaining the reaction $A_1 \rightarrow A_2$, instead of the spontaneous reaction $A_2 \rightarrow A_1$.

See Fig. 9.3.

Table 9.1 shows the analogy between the modes of the reversible thermal engine and the modes of the reversible chemical engine. If we replace

Table 9.1. Analogy between the operational modes of a thermal engine and a chemical reactor.

Thermal engine	Chemical reactor
heat engine	chemical engine
heat pump	A_1 pump
refrigerator	A_2 pump

the usual terminology 'refrigerator' by the equivalent but less common terminology 'cold pump', then the symmetry in the table is even better.

9.3 The generalized endoreversible engine

In the general structure (see Fig. 9.1(b)) we have to apply three axioms, i.e. the same three axioms as in Section 7.3:

Axiom 0: conservation of matter:

$$\Sigma N = 0 \ .$$

Axiom 1: conservation of energy:

$$\Sigma U + \Sigma W = 0 \ .$$

Axiom 2: conservation of entropy in the reversible parts:

$$\Sigma \frac{U - \mu N}{T} = 0 \ .$$

Axiom 2, i.e.

$$\frac{U_2 - \mu_4 N}{T_4} - \frac{U_1 - \mu_3 N}{T_3} = 0$$

leads to

$$\frac{U_1 - W - \mu_4 N}{T_4} = \frac{U_1 - \mu_3 N}{T_3} \ ,$$

where N denotes $N_1 = N_2$ (in accordance with Axiom 0). This yields

$$W = (1 - \frac{T_4}{T_3})U_1 + (\frac{T_4}{T_3}\mu_3 - \mu_4)N \ .$$

If we postulate

$$W = \eta U_1 + \zeta N \ ,$$

we obtain a 'vectorial efficiency' (η, ζ), with

$$\eta = 1 - \frac{T_4}{T_3} \ , \quad \zeta = \frac{T_4}{T_3}\mu_3 - \mu_4 \ .$$

If not only the engine part between reservoir (T_3, μ_3) and reservoir (T_4, μ_4) is reversible, but the whole engine is reversible, i.e. when $T_3 = T_1$, $\mu_3 = \mu_1$, $T_4 = T_2$, and $\mu_4 = \mu_2$, then η and ζ take their reversible or Carnot values, given by

$$\eta_C = 1 - \frac{T_2}{T_1} \ , \quad \zeta_C = \frac{T_2}{T_1}\mu_1 - \mu_2 \ . \tag{9.1}$$

In non-reversible conditions, the entropy creation rate is given by

$$\begin{aligned} \Sigma S(\eta, \zeta) &= \frac{U_2 - \mu_2 N}{T_2} - \frac{U_1 - \mu_1 N}{T_1} \\ &= \frac{(U_1 - W) - \mu_2 N}{T_2} - \frac{U_1 - \mu_1 N}{T_1} \\ &= \frac{U_1 - \eta U_1 - \zeta N - \mu_2 N}{T_2} - \frac{U_1 - \mu_1 N}{T_1} \ . \end{aligned}$$

Simple regrouping of terms leads to

$$\begin{aligned} \Sigma S(\eta, \zeta) &= (1 - \frac{T_2}{T_1} - \eta)\frac{U_1}{T_2} + (\frac{T_2}{T_1}\mu_1 - \mu_2 - \zeta)\frac{N}{T_2} \\ &= (\eta_C - \eta)\frac{U_1}{T_2} + (\zeta_C - \zeta)\frac{N}{T_2} \ . \end{aligned} \tag{9.2}$$

This result constitutes a 'vectorial product' generalization of eqn (3.13).

Further calculation of the functions $W(\eta, \zeta)$ and $\Sigma S(\eta, \zeta)$ will need the constitutive conduction laws for both U and N. We will not perform here such general calculations: we only need to remember that these constitutive laws have to fulfil the three inequalities (7.7)-(7.9), as well as Onsager's equality (7.10).

9.4 Two alternative generalized models ♡

As rightly remarked by Callen [2], it is instructive to recast a model, developed in terms of total energy fluxes U and particle fluxes N, into a form in terms of heat fluxes Q and particle fluxes N. For this purpose we define a heat flux Q as $U - \mu N$. The three axioms of Section 9.3 are then as follows.

Axiom 0: conservation of matter:

$$\Sigma N = 0 \ .$$

Axiom 1: conservation of energy:

$$\Sigma Q + \Sigma \mu N + \Sigma W = 0 \ .$$

Axiom 2: conservation of entropy in the reversible parts:

$$\Sigma \frac{Q}{T} = 0 \ .$$

We have to note immediately that Axiom 1 has the following consequence: we cannot speak of the 'heat current through a thermal conductor', as the heat entering the conductor at one end doesn't equal the heat leaving it at the other terminal (as soon as the two terminals are at different chemical potential μ). Therefore, we have to introduce four heat currents: Q_1, Q_2, Q_3 and Q_4, whereas two energy fluxes U_1 and U_2 sufficed in Section 9.3. See Fig. 9.4. We have $Q_3 = Q_1 + \mu_1 N - \mu_3 N$, such that the amount $(\mu_1 - \mu_3)N$ of chemical energy is degraded to heat during the transport from reservoir 1 to reservoir 3. Analogously Q_2 is larger than Q_4 by an amount $(\mu_4 - \mu_2)N$.

Axiom 2, i.e.

$$\frac{Q_4}{T_4} - \frac{Q_3}{T_3} = 0$$

now leads to

$$\frac{Q_1 + \mu_1 N - \mu_4 N - W}{T_4} = \frac{Q_1 + \mu_1 N - \mu_3 N}{T_3} \ ,$$

as well as to

$$\frac{Q_3 + \mu_3 N - \mu_4 N - W}{T_4} = \frac{Q_3}{T_3} \ .$$

This yields

$$W = (1 - \frac{T_4}{T_3})Q_1 + [\ (1 - \frac{T_4}{T_3})\mu_1 + \frac{T_4}{T_3}\mu_3 - \mu_4\ ]\ N$$

and

$$W = (1 - \frac{T_4}{T_3})Q_3 + (\mu_3 - \mu_4)N \ .$$

If we postulate

$$\text{either } W = \eta Q_1 + \vartheta N$$
$$\text{or } W = \eta Q_3 + \lambda N ,$$

we obtain either the following new 'vectorial efficiency' (η, ϑ), with

$$\eta = 1 - \frac{T_4}{T_3} \quad , \quad \vartheta = (1 - \frac{T_4}{T_3})\mu_1 + \frac{T_4}{T_3}\mu_3 - \mu_4 ,$$

with reversible values

$$\eta_C = 1 - \frac{T_2}{T_1} \quad , \quad \vartheta_C = \mu_1 - \mu_2 ,$$

or we obtain the following new 'vectorial efficiency' (η, λ), with

$$\eta = 1 - \frac{T_4}{T_3} \quad , \quad \lambda = \mu_3 - \mu_4 ,$$

with reversible values

$$\eta_C = 1 - \frac{T_2}{T_1} \quad , \quad \lambda_C = \mu_1 - \mu_2 .$$

In non-reversible conditions, we find

$$\Sigma S(\eta, \vartheta) = (\eta_C - \eta)\frac{Q_1}{T_2} + (\vartheta_C - \vartheta)\frac{N}{T_2} ,$$

or

$$\Sigma S(\eta, \lambda) = (\eta_C - \eta)\frac{Q_3}{T_2} + (\lambda_C - \lambda)\frac{N}{T_2} ,$$

analogously to eqn (9.2).

A full description based upon the fluxes Q and N is more complicated than a description based upon the fluxes U and N. Indeed, as mentioned above, we cannot speak of the heat current Q through a particular conductor g. This is caused by the fact that heat fluxes can have divergence, whereas U and N fluxes are always divergenceless. As a consequence, the notion 'constitutive law' cannot be rigorously defined for Q.

We will end the discussion of the general models, in order to concentrate upon the purely chemical case, where all temperatures are equal: $T_1 = T_2 = T_3 = T_4 = T$.

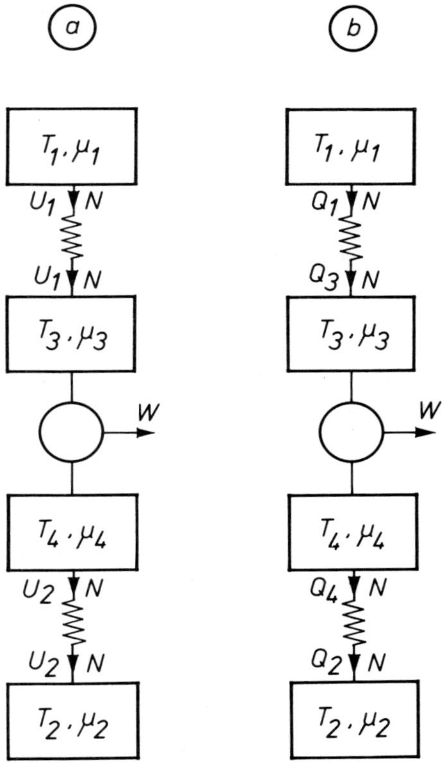

Fig. 9.4. Two thermodynamic descriptions: (a) with energy and particle currents, (b) with heat and particle currents.

9.5 The endoreversible chemical reactor

In Fig. 9.1(d) we deliberately substituted the chemical potentials μ by the more common concentrations n. Both variables can be interchanged, because of the existence of a definite relation between them, i.e. the Nernst equation:

$$\mu = \mu_o + kT \log(\frac{n}{n_o}) \ ,$$

where μ_o denotes the chemical potential at normalized concentration n_o.

We will consider the simple chemical reaction

$$A_1 - A_2 \rightleftharpoons 0 \ ,$$

where A_1 constitutes the reactant and A_2 the reaction product. Adaption of the calculation for an arbitrary reaction with arbitrary stoichiometric coefficients ν_j, i.e. for

$$\Sigma \nu_j A_j \rightleftharpoons 0 \tag{9.3}$$

is straightforward. See Appendix E.

Because we assume that the whole reactor as isothermal (at temperature T), the three general Axioms from Section 9.3 now lead to the following two simplified Axioms:

Axiom 0: conservation of matter:

$$\Sigma N = 0 \ .$$

Axiom 1: summing the conservation law of energy ($\Sigma U + \Sigma W = 0$) and $-T$ times the conservation law of entropy in the reversible parts (i.e. $-\Sigma U + \Sigma \mu N = 0$) leads to

$$\Sigma W + \Sigma \mu N = 0 \ .$$

It is clear that, because of the isothermal assumption, the two models of Section 9.4 both lead to exactly the same set of two results.

The constitutive laws for the chemical product conductors are supposed to be linear laws:

$$\begin{aligned} N &= h_1(n_1 - n_3) \\ N &= h_2(n_4 - n_2) \ , \end{aligned} \tag{9.4}$$

after Fick's particle diffusion law:

$$N = -D \nabla n \ .$$

Because of Axiom 1 we have

$$\mu_4 N + W - \mu_3 N = 0 \ . \tag{9.5}$$

If we introduce the following definition of ζ, the 'work production per mole of fuel':

$$W = \zeta N \ ,$$

then we obtain from eqn (9.5)

$$\begin{aligned} \zeta &= \mu_3 - \mu_4 \\ &= \zeta_o + kT \log(\frac{n_3}{n_4}) \ , \end{aligned}$$

where $\zeta_o = \mu_{o3} - \mu_{o4} = \mu_{o1} - \mu_{o2}$.

In the above we used the notation ζ from Sections 9.2 and 9.3. We could have used equally well ϑ or λ from Section 9.4, as all three quantities ζ, ϑ, and λ are equivalent in the isothermal case.

We now proceed analogously to the various endoreversible thermal engines we described before. Whereas for thermal engines we had to solve a set of two equations for the two unknown temperatures T_3 and T_4, we have to solve here the set

$$\begin{aligned} h_1(n_1 - n_3) &= h_2(n_4 - n_2) \\ \zeta &= \zeta_o + kT\log(\frac{n_3}{n_4}) \end{aligned}$$

for the two unknown concentrations n_3 and n_4. The solution is

$$\begin{aligned} n_3 &= \frac{h_1 n_1 + h_2 n_2}{h_1 + h_2 \exp(-\frac{\zeta - \zeta_o}{kT})} \\ n_4 &= \frac{h_1 n_1 + h_2 n_2}{h_1 \exp(\frac{\zeta - \zeta_o}{kT}) + h_2} . \end{aligned}$$

Substitution of the n_3 result into eqn (9.4) yields

$$N = \frac{n_1 - n_2 \exp(\frac{\zeta - \zeta_o}{kT})}{\frac{1}{h_1} + \frac{1}{h_2} \exp(\frac{\zeta - \zeta_o}{kT})} . \tag{9.6}$$

Figure 9.5(a) shows the function $N(\zeta)$. The curve $N(\zeta)$ is well known in chemistry and electrochemistry: it is the diffusion limited reaction rate. See, for example, [3].

This characteristic displays two different saturation currents:

- For $\zeta \to -\infty$, the reaction rate is limited to $h_1 n_1$,
- for $\zeta \to +\infty$, it is limited to $-h_2 n_2$.

It also has two different particular points:

- The intersection with the ordinate axis, as usual, is called the short-circuit point. The short-circuit current equals

$$N_{sc} = N(0) = \frac{n_1 - n_2 \exp(\frac{-\zeta_o}{kT})}{\frac{1}{h_1} + \frac{1}{h_2} \exp(\frac{-\zeta_o}{kT})} .$$

- The intersection with the abscissa axis, as usual, is called the open-circuit point. By equating expression (9.6) to zero and solving for ζ, we find the open-circuit potential:

$$\zeta_{oc} = \zeta_C = \mu_1 - \mu_2 = \zeta_o - kT\log(n_2/n_1) \ ,$$

an expression which shows some analogy with Carnot's formula $\eta_C = 1 - T_2/T_1$.

The short-circuit point and the open-circuit point together divide the $N(\zeta)$ characteristic into three parts:

- For $-\infty < \zeta < 0$, the reactor works as an A_2 pump.
- For $0 < \zeta < \zeta_{oc}$, the reactor works as a chemical engine.
- For $\zeta_{oc} < \zeta < +\infty$, the reactor works as an A_1 pump.

Multiplication of eqn (9.6) by ζ then gives

$$W = \zeta\ \frac{n_1 - n_2\exp(\frac{\zeta-\zeta_o}{kT})}{\frac{1}{h_1} + \frac{1}{h_2}\exp(\frac{\zeta-\zeta_o}{kT})}\ .$$

Finally, substitution of eqn (9.6) into

$$\Sigma S = (\zeta_C - \zeta)\frac{N}{T}$$

yields the entropy creation rate:

$$\Sigma S = [\ \frac{\zeta_o}{T} + k\log(\frac{n_1}{n_2}) - \frac{\zeta}{T}\ ]\ .\ \frac{n_1 - n_2\exp(\frac{\zeta-\zeta_o}{kT})}{\frac{1}{h_1} + \frac{1}{h_2}\exp(\frac{\zeta-\zeta_o}{kT})}\ .$$

Figure 9.5 shows the quantities $W(\zeta)$ and $\Sigma S(\zeta)$.

We see that the curve $W(\zeta)$ displays a maximum for some ζ value between 0 and ζ_C. However, unlike in the case of the Curzon–Ahlborn engine, the maximum-power condition (i.e. $dW/d\zeta = 0$) leads to a transcendental equation in the present case.

9.6 Conclusion

We have developed an analogy between the heat-diffusion-limited heat engine of Curzon and Ahlborn and the particle-diffusion-limited chemical reactor. Table 9.2 shows the analogies used.

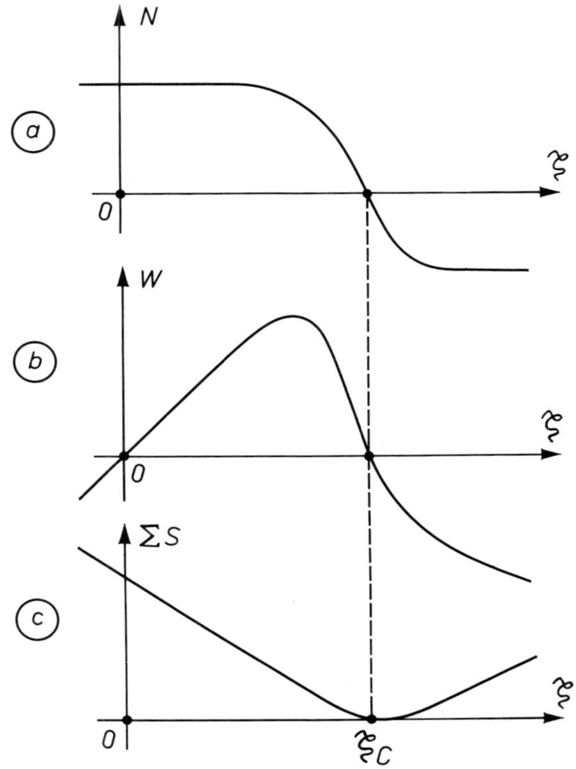

Fig. 9.5. Thermodynamic quantities as a function of the extracted molar work $\zeta = \mu_3 - \mu_4$ of an endoreversible reactor: (a) matter consumption N, (b) work production W, (c) entropy creation rate ΣS.

Table 9.2. Analogy between thermal engine and chemical reactor.

	Thermal engine	Chemical reactor	
Fourier's law	$Q = -\lambda \nabla T$	$N = -D\nabla n$	Fick's law
	$Q_1 = g_1(T_1 - T_3)$	$N = h_1(n_1 - n_3)$	
	$Q_2 = g_2(T_4 - T_2)$	$N = h_2(n_4 - n_2)$	
	$\eta = (T_3 - T_4)/T_3$	$\zeta = \mu_3 - \mu_4$	
Carnot's law	$\eta_C = (T_1 - T_2)/T_1$	$\zeta_C = \mu_1 - \mu_2$	Gibbs's law
	$W = \eta Q_1$	$W = \zeta N$	
	$\Sigma S = (\eta_C - \eta)\frac{Q_1}{T_2}$	$\Sigma S = (\zeta_C - \zeta)\frac{N}{T}$	

The resulting curves on Figs. 3.3 and 9.5 are each other's analogon. The only difference that occurs, is the following: all three curves on Fig. 3.3 show a vertical asymptote at $\eta = 1$, whereas none of the three curves on Fig. 9.5 displays a vertical asymptote. This is caused by the fact that μ_3 and μ_4 can take any (positive or negative) value, whereas T_3 and T_4 can take only positive values. Passing the frontier $\eta = 1$ would mean passing below absolute zero for either T_3 or T_4 and would need an infinite supply of work ($W \to -\infty$) and would create an infinite amount of entropy ($\Sigma S \to +\infty$).

It is clear that heat engines and chemical reactors are not the only special applications of the general endoreversible model. The reader can easily apply the scheme to other systems (pneumatic, hydraulic, etc.). The classical paper of Tolman and Fine [1] provides us with a good overview of possible applications. Table 9.3 summarizes some of these. Each time, we recognize reservoirs with a constant value of some intensive variable X and conductors which transport the associated extensive variable flow Y (i.e. extensive variable per unit of time).

If an engine model contains reservoirs of type (T, X), i.e. reservoirs at both constant temperature T and constant X, then conservation of entropy in the reversible part of the model is expressed by

$$\Sigma \frac{U - XY}{T} = 0 \ .$$

Some engines might need still more elaborate models, e.g. by considering reservoirs of type such as (T, μ, p), i.e. reservoirs which are simultaneously at constant temperature, constant chemical potential and constant pressure, instead of merely reservoirs of the type (T, μ). Conservation of entropy in the reversible part of such models is then expressed by

Table 9.3. Analogy between different thermodynamic engines.

Engine	X	Y	Transport law
heat engine	temperature T	heat flow Q	Fourier's law
chemical reactor	chem. pot. μ	particle flow N	Fick's law
pneumatic engine	pressure p	volume flow v	Poiseuille's law
electrical circuit	voltage V	current I	Ohm's law

$$\Sigma \frac{U - \Sigma_j X_j Y_j}{T} = 0 ,$$

where the summation Σ_j is performed over all intensive variables X_j, except T.

The above illustrates the power of the generalization of the concept ' endoreversible thermodynamics ' from merely ' irreversible thermodynamics with all irreversibilities caused *by heat transport* ' towards ' irreversible thermodynamics with all irreversibilities caused *by transports* '.

Bibliography

1. R. Tolman and P. Fine (1948). On the irreversible production of entropy. *Reviews of Modern Physics*, volume 20, pp. 51 - 77.

2. H. Callen (1960). *Thermodynamics.* Wiley, New York, pp. 45 - 46, 199 - 203 and 293 - 296.

3. D. Crow (1974). *Principles and applications of electrochemistry.* Chapman and Hall, London, pp. 175 - 197.

4. M. Ondrechen, R. Berry and B. Andresen (1980). Thermodynamics in finite time: a chemically driven engine. *Journal of Chemical Physics*, volume 72, pp. 5118 - 5124.

5. M. Ondrechen, B. Andresen and R. Berry (1980). Thermodynamics in finite time: processes with temperature-dependent chemical reactions. *Journal of Chemical Physics*, volume 73, pp. 5838 - 5843.

6. A. Bejan (1988). Advanced engineering thermodynamics. Wiley, New York, pp. 368 - 401.

7. A. De Vos (1991). Endoreversible thermodynamics and chemical reactions. *The Journal of Physical Chemistry*, volume 95, pp. 4534 - 4540.

10
Photosynthesis

10.1 Introduction

Photosynthesis takes a special place in the gallery of solar energy conversion systems.

Whereas

- atmospheric solar energy conversion (Chapter 4) can take place in a dead environment, and
- photothermal as well as photovoltaic conversion (Chapters 5 - 8) is created artificially,

photosynthesis cannot take place in a dead world, but does not need human intervention either. It takes an intermediate position: it 'only' needs life.

Whereas

- atmospheric solar energy conversion is global, i.e. works on a planetary scale, and
- photothermal as well as photovoltaic conversion is local, i.e. does not affect a planet as a whole,

photosynthesis can be looked at from both points of view:

- either we look at a single tree or flower, in the same way that we look at a single solar roof or a single solar cell
- or we look at the biomass as a whole, i.e. the complete plant coverage of the planet.

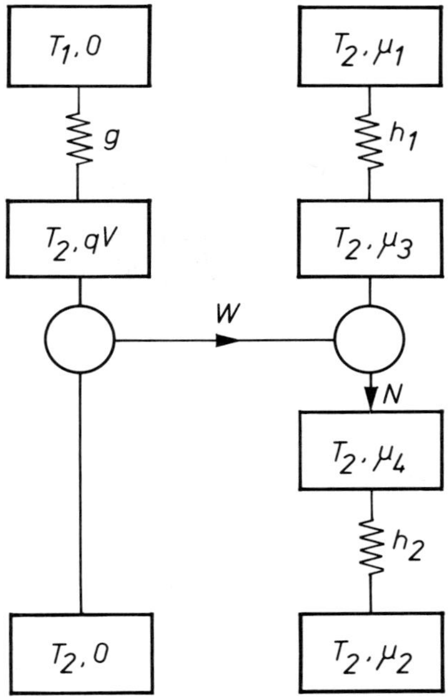

Fig. 10.1. Photosynthesis engine.

10.2 The photosynthesis engine

The photosynthesis engine is composed of two parts:

- a photovoltaic part that absorbs the solar radiation and converts it into work W, and
- a chemical reactor, which uses the work W in order to keep a chemical reaction going on in the 'reverse direction'.

See Fig. 10.1, where all reservoirs are of the type (T, μ), i.e. at constant temperature and at constant chemical potential.

In Chapter 6, we obtained the following work production rate:

$$W(V) = qV\, g\, [\int_{E_g}^{\infty} \frac{E^2 dE}{\exp(\frac{E}{kT_1}) - 1} - \int_{E_g}^{\infty} \frac{E^2 dE}{\exp(\frac{E-qV}{kT_2}) - 1}\,] \,. \tag{10.1}$$

The chemical reaction rate is given in Chapter 9:

$$N(\zeta) = \frac{n_1 - n_2 \exp(\frac{\zeta - \zeta_o}{kT_2})}{\frac{1}{h_1} + \frac{1}{h_2} \exp(\frac{\zeta - \zeta_o}{kT_2})} , \tag{10.2}$$

where ζ is $\mu_3 - \mu_4$. This equation yields a work rate $\zeta N(\zeta)$. As the work W produced by the photovoltaic part equals the work consumed by the chemical part, we have

$$\zeta N(\zeta) = -W(V) , \tag{10.3}$$

which constitutes a transcendental equation in ζ.

We define a photosynthetic efficiency

$$w = \frac{(\mu_1 - \mu_2)(-N)}{Q_s} ,$$

i.e. the ratio of the chemical energy, stored by the chemical part, to the solar energy, captured by the photovoltaic part. Note that

- the quantity $-N$ is positive,
- the quantity $(\mu_1 - \mu_2)(-N)$ is smaller than W.

Indeed, as the chemical reactor is working in the reverse direction, we have that N is negative and that the intermediate potential drop $\zeta = \mu_3 - \mu_4$ is larger than the overall potential drop $\mu_1 - \mu_2$.

Elimination of ζ and W from the equation set (10.1)-(10.2)-(10.3) finally gives rise to a relation between N and V. Carrying out such an elimination numerically yields Fig. 10.2. The curve $N(V)$ has two equal limit values:

$$\begin{aligned} N(-\infty) &= h_1 n_1 \\ N(E_g/q) &= h_1 n_1 . \end{aligned}$$

In between, it clearly displays a maximum-production point, i.e. a point where the production rate $-N$ is maximal.

Once again a short-circuit point and an open-circuit point divide the characteristic into three parts:

- For $-\infty < V < 0$, the photosynthetic reactor works as a chemical engine driving a photodiode.
- For $0 < V < V_{oc}$, the photosynthetic reactor works as an A_1 pump driven by a solar cell.
- For $V_{oc} < V < E_g/q$, the photosynthetic reactor works as a chemical engine driving a LED.

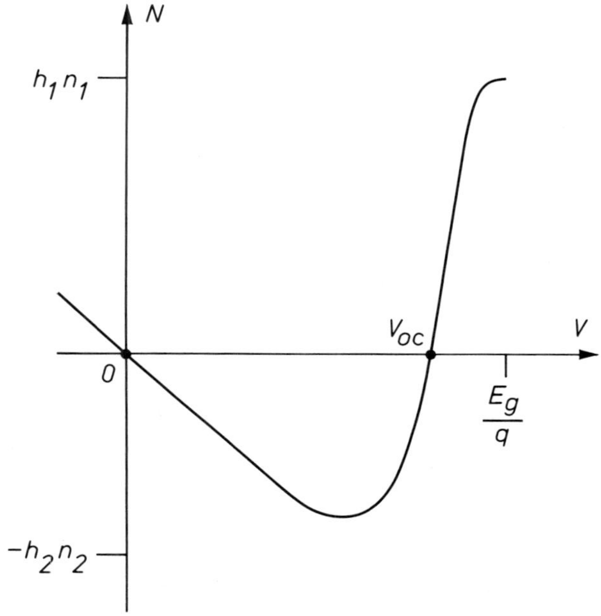

Fig. 10.2. Chemical production rate.

10.3 An example

As other planetary systems still have to be discovered, and as we are fairly sure that the Earth is the only planet of our solar system which knows life, we can give only one example of biomass: the Earth's biosphere.

Photosynthesis on the Earth is based on carbon chemistry. Photon absorption takes place mainly in the chlorophyll of the plant's leaves. With the help of this molecule, the solar energy is absorbed. Most higher plants absorb a red energy band and a blue energy band. The insensitivity to green light causes the leaves to look green.

The work produced is used to oxidize a hydrogen compound, usually water:

$$2\ H_2O \rightarrow 4\ H^+ + 4\ e^- + O_2$$

and to reduce carbon dioxide:

$$CO_2 + 4\ H^+ + 4\ e^- \rightarrow CH_2O + H_2O\ ,$$

where CH_2O is a symbolic notation for some carbohydrate. The overall reaction is

$$CO_2 + 2H_2O \rightarrow CH_2O + O_2 + H_2O\ .$$

This reaction formula can be simplified to

$$CO_2 + H_2O \rightarrow CH_2O + O_2\ , \tag{10.4}$$

but usually this is not done, in order to stress that both oxygen atoms of the oxygen gas O_2 come from the water molecules.

The Gibbs energy $\mu_{o1} - \mu_{o2}$ associated with reaction (10.4) is about 5 eV, slightly dependent upon the specific carbohydrate molecule that is synthesized. The simplest carbohydrate actually produced by plants is glucose, with formula $C_6H_{12}O_6$. This results eventually into the chemical reaction

$$6\ CO_2 + 12\ H_2O \rightarrow C_6H_{12}O_6 + 6\ O_2 + 6\ H_2O\ , \tag{10.5}$$

with $\Delta\mu_o$ equal to 29.08 eV, i.e. 4.85 eV per CH_2O group.

In order to illustrate that the role of oxygen in the water molecules is not as primordial as the role played by hydrogen, we mention here two alternative photosyntheses, taking place in purple bacteria:

$$6\ CO_2 + 12\ H_2S \rightarrow C_6H_{12}O_6 + 12\ S + 6\ H_2O\ , \tag{10.6}$$

where the role of oxygen is taken over by sulfur, and

$$6\ CO_2 + 12\ H_2 \rightarrow C_6H_{12}O_6 + 6\ H_2O\ , \tag{10.7}$$

where neither oxygen nor sulphur is necessary in the reducing agent.

Reactions (10.5), (10.6) and (10.7) play the role of the reaction

$$A_2 \rightarrow A_1$$

of Chapter 9. Indeed, we have to keep in mind that the supply of the power W makes the reaction go in the 'reverse' direction. The role of chemical compound A_2 of Chapter 9 is thus played by the reactants CO_2 and H_2O (or CO_2 and H_2S or CO_2 and H_2). The role of A_1 is played by the reaction products $C_6H_{12}O_6$ and H_2O, possibly together with O_2 or S. Thus, the photosynthetic reactor, i.e. the leaf, works as a $C_6H_{12}O_6$ pump. For sake of completeness, we have to say this only happens during insolation, since during the night, photosynthesis is replaced by respiration: carbohydrates are consumed and CO_2 is produced.

Transport of all the above chemicals is limited by the finite conductances h_1 and h_2. From Appendix E we know that of the different h_1's (associated with the different reaction products) the smallest will govern the kinetics. Analogously, the smallest of the different h_2's (associated with the different reactants) will govern the kinetics.

In most cases, kinetics are governed by the supply of the carbon dioxide. Small pores in the leaf surface, called the stomata, let the outside CO_2 enter the intercellular air space inside the leaf. These stomates thus play the role of h_2. It is fascinating to remark that the diameter of the stomate pores can be regulated: according to the actual environment (illumination, temperature, etc.), the plant changes the value of h_2. The resulting carbon dioxide concentration n_4 changes accordingly. Whereas the atmospheric CO_2 concentration n_2 is about 0.035 per cent (or 350 ppm), the intercellular concentration n_4 usually is in the range 200 to 300 ppm.

The efficiency of photosynthesis in ordinary plants is determined as follows. At the end of the growing season, the amount of produced dry plant material is weighed. The weight is converted into energy through multiplication by 15.5 kJ/g, i.e. the product of

- $\frac{1}{30\text{ g}}$, i.e. the number of moles CH_2O per unit of weight,
- 6.02×10^{23}, i.e. Avogadro's number, and
- $4.85\text{ eV} = 7.77 \times 10^{-19}\text{ J}$.

The energy, calculated this way, is then divided by the surface area of the land and by the integrated flux meter output.

In land plants, such as crop, rice and wheat, conversion efficiencies of 2 to 2.5 per cent are reported [2] and in sea plants, such as algae, efficiencies as high as 18 per cent are measured [1]. However, these numbers have to be used with caution, as biologists and agricultural engineers tend to take only the visible part of the sunlight into account for the input energy. As visible light is defined as the spectrum between $E = 1.77$ eV and $E = 3.11$ eV, we have to introduce a correction factor:

$$\frac{\int_{1.77\ eV}^{3.11\ eV} n(E)E\,dE}{\int_0^\infty n(E)E\,dE} = \frac{\int_{3.57}^{6.27} \frac{x^3\,dx}{\exp(x)-1}}{\int_0^\infty \frac{x^3\,dx}{\exp(x)-1}} = 0.368 \ ,$$

if we want to compare the above numbers with the solar energy conversion efficiencies of previous chapters.

To illustrate that photosynthesis is not only a local phenomenon, but also has to be regarded as a global phenomenon, we mention that about 0.1 per cent of the solar flux incident on the Earth is converted into chemical energy by means of photosynthesis. This figure is the more impressive, if we consider the fact that large parts of the Earth's surface are covered by oceans, by deserts and by icecaps. As an other illustration of the global effect of the Earth's biomass, Fig. 10.3 shows the CO_2 concentration, as measured in the Northern Hemisphere, during one year. The seasonal changes

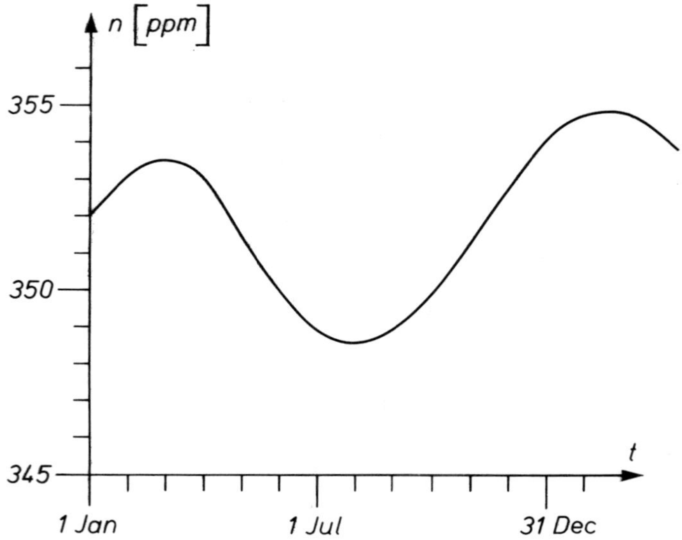

Fig. 10.3. Carbon dioxide concentration in the Earth's atmosphere.

amount to about 5 ppm and are entirely due to the seasonal changes of the activity of the vegetation: during the summer plants store carbon by photosynthesis and at other times plants release carbon as CO_2. A final illustration is the composition of the Earth's atmosphere: the 21 per cent of oxygen in the air is entirely due to photosynthesis. It is, however, true that plants already had about 2 gigayears (i.e. about 2000 million years) to produce this huge amount of oxygen gas.

Bibliography

1. J. Ryther (1959). Potential productivity of the sea. *Science*, volume 130, pp. 602 - 608.

2. J. Bonner (1962). The upper limit of crop yield. *Science*, volume **137**, pp. 11 - 15.

3. O. Bjökman, R. Pearcy, A. Harrison and H. Mooney (1972). Photosynthetic adaptation to high temperatures: a field study in Death Valley, California. *Science*, volume 175, pp. 786 - 789.

4. R. Mac Gilvery (1975). *Biochemical concepts*. Saunders Company, Philadelphia, pp. 267 - 292.

Post scriptum

If humans used solar energy on a large scale in order to perform their activities, would the planet Earth cool down? It is very important to stress that it would not! It is true that each joule we take from the solar input, does not contribute immediately to the Earth's thermal balance, but eventually this energy is dissipated by people, even if this happens on a somewhat different place and at a somewhat later instant. Thus this solar joule is eventually deposited and does contribute to the maintenance of the Earth's temperature T_p.

If, on the other hand, humans do not use solar energy for their needs, but go on burning fossil fuels and fissioning nuclear fuels, an unnatural amount of heat will be produced, thereby artificially heating the planet. Until now, humans only produce an amount of heat H of about 0.025 W/m^2, i.e. far less than the solar power $Z = (1-\rho)S = 947$ W/m^2, absorbed by the Earth. Therefore the artificial increase ΔT_p of the Earth's surface temperature is still very small and even hard to detect. Indeed:

$$4\pi R^2 \left[\frac{Z}{4} + H - (1-\gamma)\sigma(T_p + \Delta T_p)^4 \right] = 0$$

or

$$\frac{Z}{4} + H - (1-\gamma)\sigma(T_p^4 + 4T_p^3 \Delta T_p) \approx 0 \ .$$

With

$$\frac{Z}{4} - (1-\gamma)\sigma T_p^4 = 0 \tag{11.1}$$

this eventually yields

$$\Delta T_p \approx \frac{H}{Z}\, T_p \ .$$

Actually H/Z equals only about 0.00003. The corresponding heating is only

$$\Delta T_p \approx 0.01 \text{ K} \ . \tag{11.2}$$

Even if fuel consumption increases substantially, there seems to be no danger of causing climatic effects: as long as H/Z remains smaller than 0.003, ΔT_p would remain smaller than 1 K.

Unfortunately, in the above oversimplified calculation, the greenhouse coefficient γ is considered as a constant. In reality, the burning of fossil fuels causes the emission of vast amounts of CO_2 into the atmosphere. Figure 11.1 shows the carbon dioxide concentration in the Earth's atmosphere, as monitored in the period 1960 to 1990. We can recognize the seasonal variations, already mentioned in Chapter 10, but superposed on this phenomenon, we see a continuing increase of 1.2 ppm/year. This illustrates dramatically the global effect of human activity. Carbon dioxide concentration has roughly increased from about 260 ppm in the eighteenth century, i.e. before the industrial revolution, to 290 ppm in 1890 and to 350 ppm in 1990. Unfortunately, CO_2 is one of the Earth's most important greenhouse gases. Therefore an increase of CO_2 content in the atmosphere causes an increase in the greenhouse coefficient γ and a related increase in planet temperature T_p:

$$4\pi R^2 \left[\frac{Z}{4} - (1-\gamma-\Delta\gamma)\sigma(T_p+\Delta T_p)^4 \right] = 0$$

or

$$\frac{Z}{4} - (1-\gamma-\Delta\gamma)\sigma(T_p^4 + 4T_p^3\Delta T_p) \approx 0 \ .$$

Again making use of eqn (11.1), this eventually yields

$$\Delta T_p \approx \frac{\Delta\gamma}{4(1-\gamma)}\, T_p \ . \qquad (11.3)$$

An increase ΔT_p of 1 K is to be feared, as soon as $\Delta\gamma$ becomes 0.008. How $\Delta\gamma$ is related to the increase Δn of CO_2 concentration is not fully clear. It seems that $\Delta\gamma/\Delta n$ is about 0.00007 per ppm. Together with $\Delta T_p/\Delta\gamma$ about 103 K, after eqn (11.3), this results in

$$\frac{\Delta T_p}{\Delta n} \approx 0.007\ \frac{\mathrm{K}}{\mathrm{ppm}} \ ,$$

in accordance with the detailed calculations in reference [1]. Multiplying this result with $dn/dt = 1.2$ ppm/year finally yields

$$\frac{dT_p}{dt} \approx 0.01\ \mathrm{K/year} \ .$$

The reader should note that 0.01 K/year is far more disturbing than the 0.01 K of eqn (11.2). The experimental value, for the period 1881 - 1989, fortunately is somewhat smaller: $dT_p/dt = (0.45 \pm 0.15)$ K/century [6].

The safest policy is probably to keep Δn equal to zero, i.e. to use solar energy instead of fossil fuels. By the way: solar energy is free...

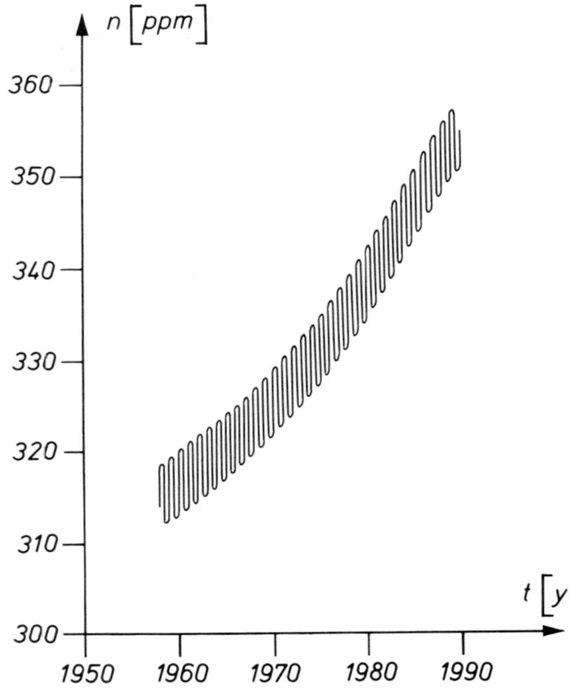

Fig. 11.1. Carbon dioxide concentration in the Earth's atmosphere.

Bibliography

1. V. Ramanathan, R. Cicerone, H. Singh and J. Kiehl (1985). Trace gas trends and their potential role in climate change. *Journal of Geophysical Research*, part D, volume 90, pp. 5547 - 5566.

2. W. Clark, T. Graedel, P. Crutzen, S. Schneider, J. la Rivière, E. Wilson, N. Keyfitz, P. Crosson, N. Rosenberg, J. Gibbons, P. Blair, H. Gwin, R. Frosch, N. Gallopoulos, J. MacNeill and W. Ruckelshaus (1989). Managing planet Earth. Special issue of the *Scientific American*, volume 261, number 3, pp. 18 - 120c.

3. R. White (1990). The great climate debate. *Scientific American*, volume 263, number 1, pp. 18 - 25.

4. P. Jones and T. Wigley (1990). Global warming trends. *Scientific American*, volume 263, number 2, pp. 66 - 73.

5. G. Davis, A. Fickett, C. Gellings, A. Lovins, R. Bevington, A. Rosenfeld, M. Ross, D. Steinmeyer, D. Bleviss, P. Walzer, A. Reddy, J. Goldemberg, W. Chandler, A. Makarov, Z. Dadi, W. Fulkerson, R. Judkins, M. Sanghvi, W. Häfele, C. Weinberg, R. Williams and J. Holdren (1990). Energy for planet Earth. Special issue of the *Scientific American*, volume 263, number 3, pp. 20 - 115.

6. J. Houghton, G. Jenkins and J. Ephraums (1990). *Climate change, the I.P.C.C. scientific assessment.* Cambridge University Press, Cambridge.

Appendix A
Constants

A.1 Mathematical constants

$\pi =$	3.14159		pi
$\gamma =$	0.57722		Euler–Mascheroni constant
$\zeta(2) =$	1.64493	$= \pi^2/6$	Riemann zeta function
$\zeta(3) =$	1.20206		
$\zeta(4) =$	1.08232	$= \pi^4/90$	
$\zeta(5) =$	1.03693		
$\zeta(6) =$	1.01734	$= \pi^6/945$	
$\zeta(7) =$	1.00835		
$\cdots$			
$\zeta'(2) =$	$-$ 0.93755		first derivative of Riemann zeta function
$\zeta'(3) =$	$-$ 0.19813		
$\zeta'(4) =$	$-$ 0.06891		
$\zeta'(5) =$	$-$ 0.02857		
$\zeta'(6) =$	$-$ 0.01285		
$\zeta'(7) =$	$-$ 0.00603		
$\cdots$			

A.2 Physical constants

$h =$	6.63×10^{-34} Js	
$' =$	4.14×10^{-15} eVs	Planck's constant
$c =$	3.00×10^{8} m/s	speed of light
$k =$	1.38×10^{-23} J/K	
$=$	8.62×10^{-5} eV/K	Boltzmann constant
$q =$	1.60×10^{-19} As	elementary charge
$\sigma =$	5.67×10^{-8} $\mathrm{Wm^{-2}K^{-4}}$	Stefan–Boltzmann constant
$\sigma' =$	1.51×10^{15} $\mathrm{s^{-1}m^{-2}K^{-3}}$	
$N_A =$	6.02×10^{23}	Avogadro's number

Appendix B
About the n-dimensional sphere ♡

The volume V of an n-dimensional sphere with radius R is proportional to R^n:

$$V = V_n R^n \ ,$$

where the proportionality coefficient V_n is the volume of the n-dimensional sphere with radius 1.

The number V_n can be found as follows. We divide the unitary hypersphere into slabs by a series of parallel hyperplanes, dx apart, as illustrated in Fig. B.1. Each section is a hypersphere of $(n-1)$ dimensions, with radius $\sqrt{1-x^2}$:

$$V_n = 2 \int_0^1 V_{n-1} \ (\ \sqrt{1-x^2} \ )^{n-1} \ dx \ .$$

Therefore we get

$$V_n = 2V_{n-1} \int_0^1 (1-x^2)^{\frac{n-1}{2}} \ dx \ .$$

Using the 'trick box' of mathematical analysis, we can prove that

$$\int_0^1 (1-x^2)^{\frac{m}{2}} \ dx = \frac{\sqrt{\pi}}{2} \ \frac{\Gamma(\frac{m}{2}+1)}{\Gamma(\frac{m}{2}+\frac{3}{2})} \ ,$$

such that we obtain the recursion relation

$$V_n = \sqrt{\pi} \ \frac{\Gamma(\frac{n}{2}+\frac{1}{2})}{\Gamma(\frac{n}{2}+1)} \ V_{n-1} \ .$$

Together with the well-known $V_2 = \pi$, this eventually yields

$$V_n = \frac{\pi^{n/2}}{\Gamma(\frac{n}{2}+1)} \ .$$

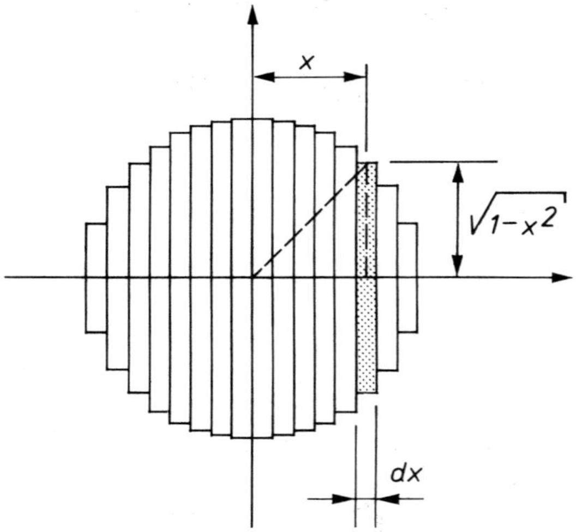

Fig. B.1. Sphere, divided into slabs.

In the above, $\Gamma(x)$ is the gamma function. For n even, we need the gamma function of an integer, which can immediately be related to the factorial function:

$$\Gamma(j) = (j-1)! = 1.2.3.\dots.(j-2).(j-1) ,$$

whereas for n odd, we need the gamma function of a half-integer number, which is related to the square root of π by

$$\Gamma(\frac{j}{2}) = \frac{1}{2}.\frac{3}{2}.\frac{5}{2}.\dots.\frac{j-4}{2}.\frac{j-2}{2}.\sqrt{\pi} .$$

The surface area S of the n-dimensional sphere is proportional to R^{n-1}:

$$S = S_n R^{n-1} ,$$

where S_n is the surface area of the hypersphere with radius 1. In order to evaluate S_n, we divide the unitary hypersphere into concentric shells, dr thick, as illustrated in Fig. B.2. An arbitrary shell has a radius r. The volume V_n of the sphere can be calculated as follows:

$$V_n = \int_0^1 S_n r^{n-1}\, dr .$$

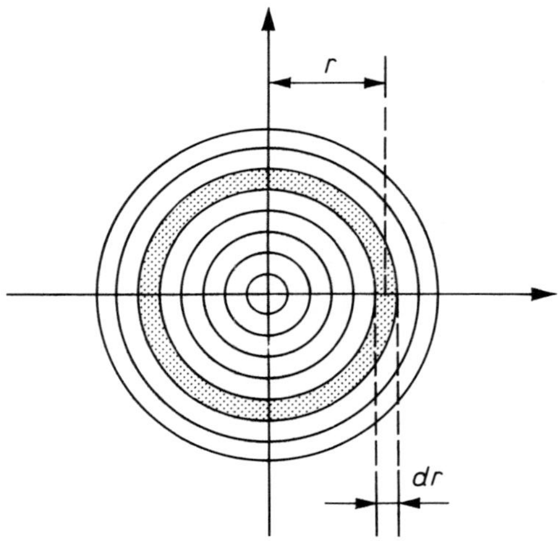

Fig. B.2. Sphere, divided into shells.

Thus

$$V_n = S_n \int_0^1 r^{n-1}\, dr \ .$$

The integral is quickly evaluated: it is equal to $1/n$, such that we get the simple result

$$S_n = nV_n \ ,$$

and hence

$$S_n = \frac{2\pi^{n/2}}{\Gamma(\frac{n}{2})} \ .$$

The surface area of the projection of the n-dimensional hypersphere on an $(n-1)$-dimensional hyperplane is nothing other than the volume of the $(n-1)$-dimensional hypersphere, i.e. $V_{n-1}R^{n-1}$. Therefore F_n, the ratio of the surface area of the n-dimensional sphere to the surface area of its projection, is given by

$$F_n = \frac{S_n}{V_{n-1}} = \frac{nV_n}{V_{n-1}} = 2\sqrt{\pi}\ \frac{\Gamma(\frac{n+1}{2})}{\Gamma(\frac{n}{2})} \ .$$

Table B.1 gives the various values of V_n, S_n and F_n. We see that F_n is alternately a rational number, and pi times a rational number. It thus is

Table B.1. Volume and surface area of unitary hypersphere.

n	V_n	S_n	F_n
1	2	2	2
2	π	$2\ \pi$	π
3	$\frac{4}{3}\ \pi$	$4\ \pi$	4
4	$\frac{1}{2}\ \pi^2$	$2\ \pi^2$	$\frac{3}{2}\ \pi$
5	$\frac{8}{15}\ \pi^2$	$\frac{8}{3}\ \pi^2$	$\frac{16}{3}$
6	$\frac{1}{6}\ \pi^3$	π^3	$\frac{15}{8}\ \pi$
7	$\frac{16}{105}\ \pi^3$	$\frac{16}{15}\ \pi^3$	$\frac{32}{5}$
8	$\frac{1}{24}\ \pi^4$	$\frac{1}{3}\ \pi^4$	$\frac{35}{16}\ \pi$
9	$\frac{32}{945}\ \pi^4$	$\frac{32}{105}\ \pi^4$	$\frac{256}{35}$
10	$\frac{1}{120}\ \pi^5$	$\frac{1}{12}\ \pi^5$	$\frac{315}{128}\ \pi$
11	$\frac{64}{10395}\ \pi^5$	$\frac{64}{945}\ \pi^5$	$\frac{512}{63}$
12	$\frac{1}{720}\ \pi^6$	$\frac{1}{60}\ \pi^6$	$\frac{693}{256}\ \pi$
...			
25	$\frac{8192}{7\,905\,853\,580\,625}\ \pi^{12}$	$\frac{8192}{316\,234\,143\,225}\ \pi^{12}$	$\frac{8\,388\,608}{676\,039}$
...			

alternately rational and transcendental. For n sufficiently large, we have the following asymptotic approximation:

$$F_n \approx \sqrt{2\pi n}\ .$$

Appendix C
Kepler's second law ♡

Kepler's first law says that a planet describes an elliptic orbit around the Sun. His second law says that in equal times, equal sectors of the ellipse are covered. Thus

$$\frac{dS}{dt} = c \,, \tag{C.1}$$

where S denotes the surface area of the covered sector and c denotes a constant. See Fig. C.1.

Equation (C.1) can be written differently:

$$dS = c\, dt \,. \tag{C.2}$$

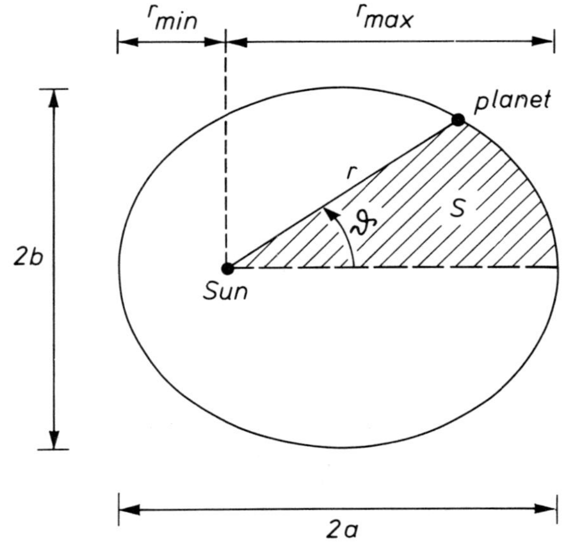

Fig. C.1. Orbit of a planet around the Sun.

By integration over one revolution period τ, this becomes:

$$S = c\tau \ .$$

As the surface area of an ellipse with major axis a and minor axis b equals πab, we eventually find an expression for the constant c in terms of τ:

$$c = \frac{\pi ab}{\tau} \ .$$

Taking this into account as well as

$$dS = \frac{1}{2} r^2 d\vartheta \ ,$$

eqn (C.2) becomes

$$\frac{1}{2} r^2 d\vartheta = \frac{\pi ab}{\tau} \, dt \ ,$$

such that

$$\frac{1}{\tau} \oint \frac{dt}{r^2} = \frac{1}{2\pi ab} \oint d\vartheta \ .$$

Because

$$\oint d\vartheta = 2\pi \ ,$$

this becomes

$$\frac{1}{\tau} \oint \frac{dt}{r^2} = \frac{1}{ab} \ .$$

This result means that the time average of $1/r^2$ over one revolution equals $1/ab$. In other words, the ellipse with half axes a and b has the same time average of $1/r^2$ as the circle with radius $\sqrt{ab}$. This quantity $\sqrt{ab}$ should not be confused with the quantity 'average distance' Sun–planet, tabulated in most astrophysics textbooks and defined as

$$\frac{r_{min} + r_{max}}{2} \ ,$$

a quantity exactly equal to a, as can be verified either by inspection of Fig. C.1 or by straightforward calculation from

$$\begin{aligned} r_{min} &= \frac{b^2}{a + \sqrt{a^2 - b^2}} \\ r_{max} &= \frac{b^2}{a - \sqrt{a^2 - b^2}} \ . \end{aligned}$$

Table C.1 shows that indeed, for some planets, $\sqrt{ab}$ is somewhat smaller than a.

Table C.1. Distance of the various planets from the Sun.

Planet	$\sqrt{ab}$ (Gm)	a (Gm)	b (Gm)	r_{min} (Gm)	r_{max} (Gm)
Mercury	57	58	57	46	70
Venus	108	108	108	107	109
Earth	150	150	150	147	152
Mars	227	228	227	206	249
Jupiter	778	778	777	741	816
Saturn	1426	1427	1425	1348	1506
Uranus	2868	2870	2866	2734	3005
Neptune	4497	4497	4496	4458	4535
Pluto	5820	5910	5720	4430	7390

Appendix D
About a fourth-degree equation ♡

If we use here the notation x for $1 - \eta$ in eqn (4.11), we obtain the quartic equation

$$x^4 + 4x - 3 = 0 \ .$$

This equation has one real solution satisfying

$$0 < x < 1 \ .$$

That solution has the following analytical form:

$$x = \sqrt{\frac{1}{\sqrt{c}} - c} \ - \ \sqrt{c} \ ,$$

where c is the only real solution of the cubic equation

$$4c^3 + 3c - 1 = 0 \ ,$$

i.e.

$$c = \frac{1}{2} \left(\sqrt[3]{\sqrt{2} + 1} \ - \ \sqrt[3]{\sqrt{2} - 1} \right) .$$

Thus x is an irrational number which mathematicians call a nested radical, i.e. a number written with roots within roots. More precisely, x is a triply nested radical, as it contains a square root within a cube root within a square root within a square root. We have

$$\begin{aligned} c &= 0.29804 \\ x &= 0.69251 \ . \end{aligned}$$

In order to evaluate expression (4.10), we have to substitute the solution x into

$$y = \frac{(1-x)x^3}{1+x^3} \ ,$$

which yields

$$y = 1 - \frac{4}{3} \left(\sqrt{\frac{1}{\sqrt{c}} - c} - \sqrt{c} \right) .$$

We have

$$y = 0.07666 .$$

In order to evaluate expression (4.16), we have to substitute the solution x into

$$z = \frac{(1 - x)x^3}{1 + x^4} ,$$

which yields

$$z = \frac{3}{4} \frac{1}{\sqrt{\frac{1}{\sqrt{c}} - c} - \sqrt{c}} - 1 .$$

We have

$$z = 0.08303 .$$

Appendix E
The general chemical reaction ♡

To avoid making notations too cumbersome, we rewrite the chemical reaction (9.3) as

$$aA + bB + cC + \ldots \rightleftharpoons pP + qQ + rR + \ldots \,.$$

The stoichiometric numbers a, b, c, ... and p, q, r, ... are assumed to be positive integers, so that A, B, C, ... can be called the reactants and P, Q, R, ... the products. See Fig. E.1.

If we denote by N_a, N_b, N_c, ..., the number of molecules A, B, C, ... annihilated per unit of time and by N_p, N_q, N_r, ... the number of molecules P, Q, R, ... created per unit of time, then Axiom 0, i.e. conservation of matter, can be written as follows:

$$\frac{N_a}{a} = \frac{N_b}{b} = \frac{N_c}{c} = \ldots = \frac{N_p}{p} = \frac{N_q}{q} = \frac{N_r}{r} = \ldots \,.$$

For convenience we give this quantity the notation N.

Axiom 1 leads to

$$W - \mu_{3a}N_a - \mu_{3b}N_b - \mu_{3c}N_c - \ldots + \mu_{4p}N_p + \mu_{4q}N_q + \mu_{4r}N_r + \ldots = 0$$

or

$$W - (a\mu_{3a} + b\mu_{3b} + c\mu_{3c} + \ldots - p\mu_{4p} - q\mu_{4q} - r\mu_{4r} - \ldots)N = 0 \,,$$

i.e.

$$\zeta = a\mu_{3a} + b\mu_{3b} + c\mu_{3c} + \ldots - p\mu_{4p} - q\mu_{4q} - r\mu_{4r} - \ldots \,.$$

The quantity $a\mu_a + b\mu_b + c\mu_c + \ldots - p\mu_p - q\mu_q - r\mu_r - \ldots$ is sometimes called the de Donder affinity. Taking now the Nernst law into account, the expression for ζ becomes

$$\zeta = \zeta_o + kT \, \log \left(\frac{n_{3a}^a \, n_{3b}^b \, n_{3c}^c \, \cdots}{n_{4p}^p \, n_{4q}^q \, n_{4r}^r \, \cdots} \, n_o^{p+q+r+\ldots-a-b-c-\ldots} \right) . \tag{E.1}$$

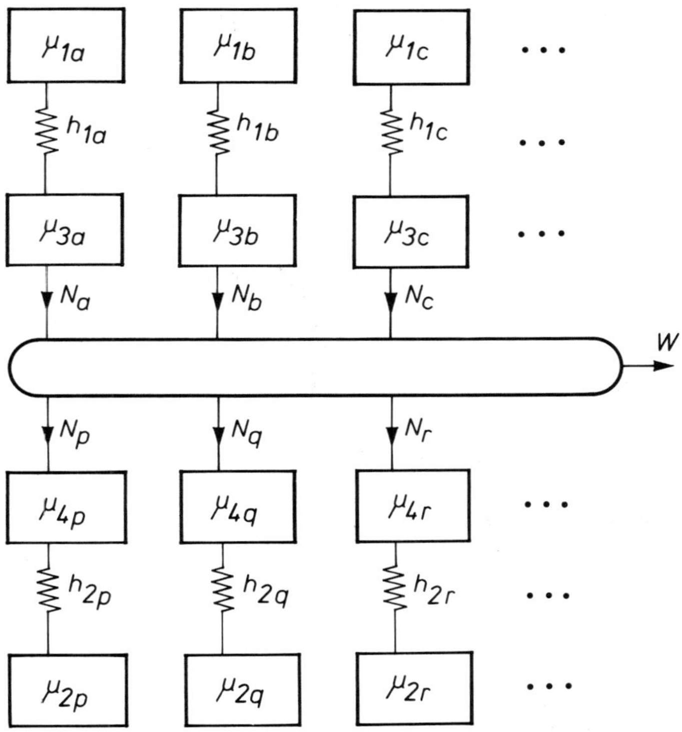

Fig. E.1. General chemical reactor.

Finally, the transport equations are:

$$\begin{aligned} N_a &= h_{1a}(n_{1a} - n_{3a}) \\ N_b &= h_{1b}(n_{1b} - n_{3b}) \\ N_c &= h_{1c}(n_{1c} - n_{3c}) \\ &\dots \end{aligned} \tag{E.2}$$

and

$$\begin{aligned} N_p &= h_{2p}(n_{4p} - n_{2p}) \\ N_q &= h_{2q}(n_{4q} - n_{2q}) \\ N_r &= h_{2r}(n_{4r} - n_{2r}) \\ &\dots \quad . \end{aligned} \tag{E.3}$$

We solve eqns (E.2) and (E.3) for n_{3a}, n_{3b}, n_{3c}, ..., n_{4p}, n_{4q}, n_{4r}, ... and substitute the result into eqn (E.1). This yields:

$$\zeta =, \zeta_o + $$
$$kT \log [\frac{(n_{1a} - \frac{N_a}{h_{1a}})^a (n_{1b} - \frac{N_b}{h_{1b}})^b (n_{1c} - \frac{N_c}{h_{1c}})^c \ldots}{(n_{2p} + \frac{N_p}{h_{2p}})^p (n_{2q} + \frac{N_q}{h_{2q}})^q (n_{2r} + \frac{N_r}{h_{2r}})^r \ldots} n_o^{p+q+r+\ldots-a-b-c-\ldots}]$$

or

$$\exp(\frac{\zeta - \zeta_o}{kT}) =$$
$$\frac{(n_{1a} - \frac{N}{ah_{1a}})^a (n_{1b} - \frac{N}{bh_{1b}})^b (n_{1c} - \frac{N}{ch_{1c}})^c \ldots}{(n_{2p} + \frac{N}{ph_{2p}})^p (n_{2q} + \frac{N}{qh_{2q}})^q (n_{2r} + \frac{N}{rh_{2r}})^r \ldots} n_o^{p+q+r+\ldots-a-b-c-\ldots} .$$

This leads to a polynomial equation in N:

$$\exp(\frac{\zeta - \zeta_o}{kT}) \; (n_{2p} + \frac{N}{ph_{2p}})^p (n_{2q} + \frac{N}{qh_{2q}})^q (n_{2r} + \frac{N}{rh_{2r}})^r \ldots$$
$$= \; n_o^{p+q+r+\ldots-a-b-c-\ldots} \; (n_{1a} - \frac{N}{ah_{1a}})^a (n_{1b} - \frac{N}{bh_{1b}})^b (n_{1c} - \frac{N}{ch_{1c}})^c \ldots . \tag{E.4}$$

The degree of this equation equals

$$\max(a + b + c + \ldots \;,\; p + q + r + \ldots) .$$

Solution of the equation gives the desired characteristic $N(\zeta)$.

It often happens that the equation yields more than one real solution. This gives rise to several branches $N(\zeta)$, as illustrated by an example in Fig. E.2. Of course only the branch that intersects the abscissa is physically relevant. This branch has two horizontal asymptotes:

$$\lim_{\zeta \to -\infty} N \;=\; \min(ah_{1a}n_{1a} \;,\; bh_{1b}n_{1b} \;,\; ch_{1c}n_{1c} \;,\; \ldots)$$
$$\lim_{\zeta \to +\infty} N \;=-\; \min(ph_{2p}n_{2p} \;,\; qh_{2q}n_{2q} \;,\; rh_{2r}n_{2r} \;,\; \ldots) .$$

This can be seen as follows. For $\zeta \to -\infty$, eqn (E.4) becomes

$$(n_{1a} - \frac{N}{ah_{1a}})^a (n_{1b} - \frac{N}{bh_{1b}})^b (n_{1c} - \frac{N}{ch_{1c}})^c \ldots = 0 ,$$

a polynomial equation in N with obvious roots

$$ah_{1a}n_{1a}\ ,\ \ bh_{1b}n_{1b}\ ,\ \ ch_{1c}n_{1c}\ ,\ \ \ldots\ .$$

The smallest of these roots is the physically relevant $N(-\infty)$. Analogously, $\zeta \to +\infty$ yields the polynomial equation

$$(n_{2p} + \frac{N}{ph_{2p}})^p (n_{2q} + \frac{N}{qh_{2q}})^q (n_{2r} + \frac{N}{rh_{2r}})^r \ldots = 0\ ,$$

with roots

$$-ph_{2p}n_{2p}\ ,\ \ -qh_{2q}n_{2q}\ ,\ \ -rh_{2r}n_{2r}\ ,\ \ \ldots\ ,$$

where again the value closest to zero has to be retained.

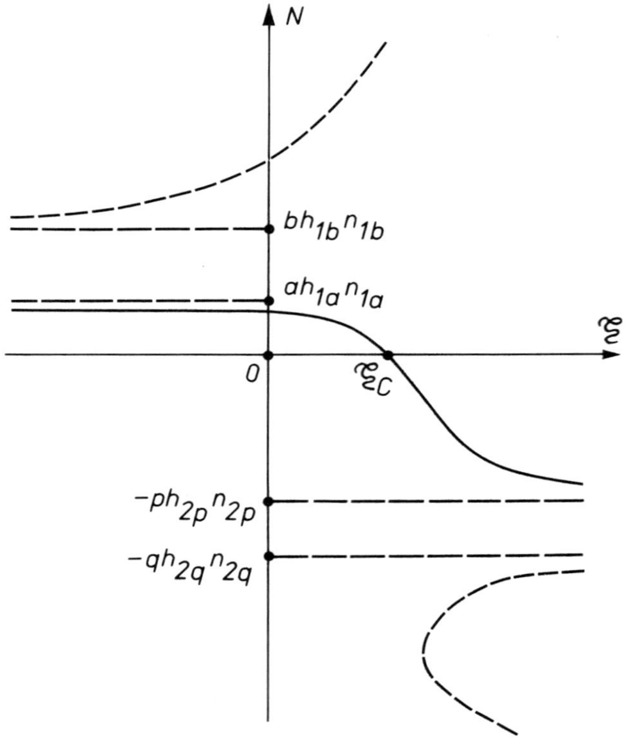

Fig. E.2. Reaction rate characteristic $N(\zeta)$ of chemical reaction aA + bB → pP + qQ, with transport conductances chosen such that $ah_{1a}n_{1a} < bh_{1b}n_{1b}$ and $ph_{2p}n_{2p} < qh_{2q}n_{2q}$ (unbroken curve: physically relevant branch; dashed curves: physically irrelevant branches).

Appendix F
List of symbols

Symbol	Meaning	Unit
a	ellipse's half major axis	m
A_1	projected equatorial surface area	m^2
A_2	projected polar surface area	m^2
A_3	equatorial surface area	m^2
A_4	polar surface area	m^2
b	ellipse's half minor axis	m
c	speed of light	ms^{-1}
C	concentration factor	1
d	thickness	m
D	diffusion coefficient	m^2s^{-1}
E	energy	J
E_F	Fermi energy	J
E_{Fc}	Fermi energy of conduction band	J
E_{Fv}	Fermi energy of valence band	J
E_{F1}	Fermi energy of level 1	J
E_{F2}	Fermi energy of level 2	J
E_g	energy gap	J
E_{gi}	i^{th} energy gap	J
E_1	energy level 1	J
E_2	energy level 2	J

Symbol	Meaning	Unit
f	occupation probability	1
f	sunlight dilution factor	1
f_1	occupation probability of level 1	1
f_2	occupation probability of level 2	1
$f_\star$	starlight dilution factor	1
$f_\oslash$	moonlight dilution factor	1
F_n	n-dimensional projection factor	1
g	number of modes	1
g	heat conductance	(*)
g_1	heat conductance from reservoir 1	(*)
g_2	heat conductance to reservoir 2	(*)
h	spectral photon flux	$m^{-2}s^{-1}J^{-1}$
h	Planck's constant	Js
h_1	particle conductance from reservoir 1	m^3s^{-1}
h_2	particle conductance to reservoir 2	m^3s^{-1}
H	human energy dissipation density	Wm^{-2}
i	label of stack element	1
I	electric current	A
I_i	i^{th} electric current	A
I_{sat}	saturation electric current	A
I_{sc}	short-circuit electric current	A
k	Boltzmann's constant	JK^{-1}

(*) The unit of g, g_1 and g_2 depends on the particular endoreversible engine under consideration.

Symbol	Meaning	Unit
n	number of spatial dimensions	1
n	number of elements in stack	1
n	particle concentration	m^{-3}
n_o	reference particle concentration	m^{-3}
n_1	particle concentration of reservoir 1	m^{-3}
n_2	particle concentration of reservoir 2	m^{-3}
n_3	particle concentration of reservoir 3	m^{-3}
n_4	particle concentration of reservoir 4	m^{-3}
N	particle current	s^{-1}
p	momentum	Jsm^{-1}
p_1	momentum coordinate	Jsm^{-1}
p_2	momentum coordinate	Jsm^{-1}
p_3	momentum coordinate	Jsm^{-1}
q	elementary charge	As
Q	heat current	W
Q_i	i^{th} heat current	W
Q_{in}	incident heat current	W
Q_s	solar energy current	W
Q_{sat}	saturation heat current	W
Q_{sc}	short-circuit heat current	W
Q_1	heat current supplied by reservoir 1	W
Q_2	heat current drained by reservoir 2	W
Q_3	heat current reaching reservoir 3	W
$Q_{\oslash}$	moon radiation density	Wm^{-2}
r	distance Sun–planet	m
r_{max}	maximum distance Sun–planet	m
r_{min}	minimum distance Sun–planet	m
$r_{\oslash}$	distance moon–planet	m
R	planet radius	m
R_s	Sun radius	m
$R_{\oslash}$	moon radius	m

Symbol	Meaning	Unit
s	spin	1
S	surface area	m^2
S	solar constant	Wm^{-2}
S	entropy production rate	WK^{-1}
S_n	surface area of n-dimensional unit sphere	1
t	time	s
T	temperature	K
T_{exp}	experimental surface temperature	K
T_i	i^{th} temperature	K
T_p	planet temperature	K
T_s	Sun temperature	K
T_0	empty space temperature	K
T_1	temperature of reservoir 1	K
T_2	temperature of reservoir 2	K
T_3	temperature of reservoir 3	K
T_4	temperature of reservoir 4	K
$T_\star$	star temperature	K
U	energy current	W
V	volume	m^3
V	voltage	V
V_i	i^{th} voltage	V
V_n	volume of n-dimensional unit sphere	1
V_{oc}	open-circuit voltage	V
V_1	voltage of reservoir 1	V
V_2	voltage of reservoir 2	V
V_3	voltage of reservoir 3	V
w	solar energy conversion efficiency	1
w_i	i^{th} solar energy conversion efficiency	1
W	work current	W
W_i	i^{th} work current	W

Symbol	Meaning	Unit
x	minus inverse temperature	K^{-1}
x_g	dimensionless energy gap	1
x_1	minus inverse temperature of reservoir 1	K^{-1}
x_1	space coordinate	m
x_2	minus inverse temperature of reservoir 2	K^{-1}
x_2	space coordinate	m
x_3	minus inverse temperature of reservoir 3	K^{-1}
x_3	space coordinate	m
x_4	minus inverse temperature of reservoir 4	K^{-1}
X	energy current	W
X	intensive variable	(*)
y	affinity μ/T	JK^{-1}
y_1	affinity μ/T of reservoir 1	JK^{-1}
y_2	affinity μ/T of reservoir 2	JK^{-1}
y_3	affinity μ/T of reservoir 3	JK^{-1}
y_4	affinity μ/T of reservoir 4	JK^{-1}
Y	particle current	s^{-1}
Y	extensive variable current	(*)
Z	absorbed part of solar constant	Wm^{-2}

(*) The units of X and Y depend on the particular variable couple under consideration.

Symbol	Meaning	Unit
α	dissipation distribution	1
α	absorptivity	1
γ	Euler–Mascheroni constant	1
γ	planet infrared reflectivity	1
Γ	gamma function	1
ϵ	emissivity	1
ϵ_i	i^{th} emissivity	1
ζ	Riemann zeta function	1
ζ	dimensionless voltage	1
ζ	chemical reactor potential	J
ζ'	first derivative of Riemann zeta function	1
ζ_C	Carnot value of dimensionless voltage	1
ζ_C	Carnot value of chemical reactor potential	J
ζ_{eq}	equilibrium dimensionless voltage	1
ζ_o	chemical reactor reference potential	J
ζ_{oc}	open-circuit dimensionless voltage	1
ζ_{oc}	open-circuit chemical reactor potential	J
η	engine efficiency	1
η_C	Carnot value of efficiency	1
$\eta_{C\&A}$	Curzon–Ahlborn value of efficiency	1
η_{opt}	optimal efficiency	1
ϑ	angle of incidence	1
ϑ	dimensionless chemical potential	1
ϑ_C	Carnot value of dimensionless chemical potential	1
λ	thermal conductivity	$\mathrm{Wm^{-1}K^{-1}}$
λ	dimensionless chemical potential	1
λ_C	Carnot value of dimensionless chemical potential	1

Symbol	Meaning	Unit
μ	chemical potential	J
μ_o	reference chemical potential	J
μ_1	chemical potential of reservoir 1	J
μ_2	chemical potential of reservoir 2	J
μ_3	chemical potential of reservoir 3	J
μ_4	chemical potential of reservoir 4	J
ν_j	stoichiometric coefficient	1
π	pi	1
ρ	planet albedo	1
σ	Stefan–Boltzmann constant	$\mathrm{Wm^{-2}K^{-4}}$
σ_n	n-dimensional Stefan–Boltzmann constant	$\mathrm{Wm^{-(n-1)}K^{-(n+1)}}$
σ'	radiation constant	$\mathrm{m^{-2}K^{-3}}$
σ'_n	n-dimensional radiation constant	$\mathrm{m^{-(n-1)}K^{-n}}$
Σ	integration over a closed surface	1
τ	oscillation period	s
φ	zone boundary latitude	1
ω	solid angle	1
$\heartsuit$	section for special interest	–

Index

V

W

Z